# Grundlehren der mathematischen Wissenschaften 214

*A Series of Comprehensive Studies in Mathematics*

B. L. van der Waerden

# Group Theory and Quantum Mechanics

Corrected Printing

With 10 Figures

Springer-Verlag
Berlin Heidelberg New York 1980

Prof. Dr. B. L. van der Waerden
Mathematisches Institut der Universität Zürich

Translation of the German Original Edition: Die Grundlehren der mathematischen
Wissenschaften Band 37, Die Gruppentheoretische Methode in der Quantenmechanik. Publisher: Verlag von Julius Springer, Berlin 1932


AMS Subject Classification (1970) 81–02, 81 A 09, 81 A 78


ISBN-13:978-3-642-65862-4      e-ISBN-13:978-3-642-65860-0
DOI: 10.1007/978-3-642-65860-0

Library of Congress Cataloging in Publication Data
Waerden, Bartel Leenert van der, 1903.      Group theory and quantum mechanics.
(Die Grundlehren der mathematischen Wissenschaften in Einzeldarstellungen mit besonderer
Berücksichtigung der Anwendungsgebiete, Bd. 214) Translation of Die gruppentheoretische
Methode in der Quantenmechanik.
Bibliography: p.
1. Quantum theory. 2. Groups, Theory of. I. Title. II. Series: Die Grundlehren
der mathematischen Wissenschaften in Einzeldarstellungen, Bd. 214)
QC174.17.G7W313      530.1'2      74—13914

2140/3140-543210

# Preface

The German edition of this book appeared in 1932 under the title "Die gruppentheoretische Methode in der Quantenmechanik". Its aim was, to explain the fundamental notions of the Theory of Groups and their Representations, and the application of this theory to the Quantum Mechanics of Atoms and Molecules. The book was mainly written for the benefit of physicists who were supposed to be familiar with Quantum Mechanics. However, it turned out that it was also used by mathematicians who wanted to learn Quantum Mechanics from it. Naturally, the physical parts were too difficult for mathematicians, whereas the mathematical parts were sometimes too difficult for physicists. The German language created an additional difficulty for many readers.

In order to make the book more readable for physicists and mathematicians alike, I have rewritten the whole volume. The changes are most notable in Chapters 1 and 6. In Chapter 1, I have tried to give a mathematically rigorous exposition of the principles of Quantum Mechanics. This was possible because recent investigations in the theory of self-adjoint linear operators have made the mathematical foundation of Quantum Mechanics much clearer than it was in 1932.

Chapter 6, on Molecule Spectra, was too much condensed in the German edition. I hope it is now easier to understand.

In Chapter 2–5 too, numerous changes were made in order to make the book more readable and more useful.

B. L. VAN DER WAERDEN

Zurich, February 1974

# Table of Contents

Chapter I

# Fundamental Notions of Quantum Mechanics

## § 1. Wave Functions

According to Wave Mechanics, a *pure state*[1] of a mechanical system is
defined at any time by a wave function $\Psi$. The mechanical systems considered in this book are systems such as atoms or molecules, each consisting of a finite number of particles (electrons and nuclei). The wave
function $\Psi$ is a complex-valued function of the coordinates of the
particles, dependent on time, which is supposed to satisfy Schrödinger's
equation

$$(1.1) \qquad H\Psi + \frac{\hbar}{i} \frac{\partial \Psi}{\partial t} = 0.$$

As long as spin is neglected, the coordinates occurring in $\Psi$ are the
orthogonal coordinates $q_g$ or $x_g$, $y_g$, $z_g$ of the $f$ particles; the index $g$
goes from 1 to $f$. *The Hamiltonian* or *energy operator H is defined as
follows:* Let $\mu = \mu_g$ be the mass of any one of the particles. The classical
Hamiltonian is an expression of the form

$$(1.2) \qquad T + U = \sum_g \frac{1}{2\mu} (p_x^2 + p_y^2 + p_z^2) + U(q),$$

$U(q)$ being the potential energy function. In this classical expression, one
has to replace the momentum components $p_x, p_y, p_z$ by differential
operators

$$\frac{\hbar}{i} \frac{\partial}{\partial x}, \quad \frac{\hbar}{i} \frac{\partial}{\partial y}, \quad \frac{\hbar}{i} \frac{\partial}{\partial z},$$

*thus obtaining the operator*

$$H = \sum_g -\frac{\hbar^2}{2\mu} \left( \frac{\partial^2}{\partial x^2} + \frac{\partial^2}{\partial y^2} + \frac{\partial^2}{\partial z^2} \right) + U(q)$$

$$= \sum_g -\frac{\hbar^2}{2\mu_g} \Delta_g + U(q)$$

---

[1] The notion "pure state", as opposed to "mixture", is due to J. von Neumann. See e.g.
his "Principles of Quantum Mechanics".

The wave functions $\Psi$ are supposed to be integrable in the sense of Lebesgue[2] and to have a finite square integral over $q$-space

$$(1.3) \qquad \langle \Psi, \Psi \rangle = \int \Psi^* \Psi \, dq$$

(* means complex conjugate). The functions $\Psi$ satisfying these conditions form a *Hilbert space*. This notion will be more fully discussed in § 2.

If the integral (1.3) is zero, the function $\Psi$ is almost everywhere zero[2] and defines no state. If $\langle \Psi, \Psi \rangle$ is not zero, we can multiply the function $\Psi$ by a constant factor $\lambda$ so that $\langle \Psi, \Psi \rangle$ becomes 1; the function is then "normed". The functions $\Psi$ and $\lambda \Psi$ define the same state. If $\Psi$ is normed, the integral of $\Psi^* \Psi$ over any measurable domain $D$ in $q$-space is the probability for the system of particles to be found in $D$. This is Born's statistical interpretation of Wave Mechanics.

Most important are the *stationary states* of the system, i.e. states which depend on time only by a factor $e^{-i\omega t}$:

$$(1.4) \qquad \Psi = \psi(q) \, e^{-i\omega t}$$

If $\Psi$ is to satisfy (1.1), $\psi$ must satisfy the equation

$$(1.5) \qquad H\psi = E\psi \quad \text{with } E = \hbar\omega.$$

The Eq. (1.5) defines an eigenvalue-problem. The unknowns of the problem are the *eigenfunction* $\psi$ and the *eigenvalue* $E$. Eigenfunctions having a finite square integral are possible only if $E$ belongs to the *point spectrum* of the operator $H$. In the continuous spectrum solutions of the form (1.4) behave at large distances of the particles like plane waves, and the integral (1.3) becomes infinite. However, functions (1.4) belonging to different frequencies $\omega$ may be combined by integration to form *wave packets* having a finite integral (1.3).

To illustrate this, let us consider the case of a free particle. In this case, the energy operator is

$$(1.6) \qquad H = -\frac{\hbar^2}{2\mu} \left( \frac{\partial^2}{\partial x^2} + \frac{\partial^2}{\partial y^2} + \frac{\partial^2}{\partial z^2} \right)$$

and the solutions of (1.5) are plane waves

$$(1.7) \qquad \psi = e^{i(kq)}; \quad (kq) = k_1 x + k_2 y + k_3 z$$

---

[2] H. Lebesgue: Leçons sur l'intégration. Paris: 1904.

with

$$E = \frac{\hbar^2}{2\mu}(kk).$$

Now a function like (1.7) clearly has an infinite square integral, but by integration over $k$-space we can form wave packets having a finite square integral as follows:

$$(1.8) \qquad \psi(q) = (2\pi)^{-\frac{3}{2}} \int\int\int \varphi(k)\, e^{i(kq)}\, dk_1\, dk_2\, dk_3 \, ,$$

the triple integral over the whole $k$-space being defined as a "limit in the mean" of an integral over a finite portion of $k$-space[3]. In fact, Plancherel's theorem says that to every function $\varphi(k)$ in $k$-space having a finite square integral corresponds a function $\psi(q)$ in $q$-space having just the same square integral, and vice versa. The solution of (1.8) is

$$(1.9) \qquad \varphi(k) = (2\pi)^{-\frac{3}{2}} \int\int\int \psi(q)\, e^{-i(kq)}\, dx\, dy\, dz \, ,$$

the integral over the $q$-space being defined as before as a limit in the mean.

In this book, we shall mainly be concerned with states belonging to the point spectrum. For these states, the integral (1.3) is finite and the energy $H$ has a quite definite value $E$. More generally, every measurable quantity (such as a momentum or moment of momentum) is represented in quantum mechanics by a linear operator $A$ operating on $\Psi$, and if $\Psi$ is an eigenfunction of $A$ with eigenvalue $\lambda$, the quantity $A$ is supposed to have in the state $\Psi$ the value $\lambda$.

By Bohr's well-known Frequency Formula, the energy-values $E$ determine the visible spectrum of the atom or molecule. By the influence of the radiation field, the atom may jump from a state of higher energy $E_1$ to a state of lower energy $E_2$, or conversely, and the frequency $\omega' = 2\pi\nu'$ of the emitted or absorbed light is given by

$$(1.10) \qquad E_1 - E_2 = h\nu' = \hbar\omega' \, .$$

The energy values $E$ are also called *terms*. For spectroscopic purposes it is convenient to divide the energies by $hc$, thus obtaining *wave numbers* or inverse wave lengths, expressed in $\mathrm{cm}^{-1}$:

$$\frac{1}{\lambda} = \frac{\omega}{2\pi c} = \frac{E}{2\pi\hbar c} = \frac{E}{hc} \, .$$

---

[3] A function $\psi = \psi(q)$ is called "limit in the mean" (l.i.m.) of a sequence of functions $\psi_n$, if the integral $\int |\psi_n - \psi|^2\, dq$ converges to zero for $n \to \infty$.

## § 2. Hilbert Spaces

The notion "Hilbert space" may be defined in several ways. For our purpose, the simplest way is to start with functions $\varphi$ of the variables $q$, integrable in the sense of Lebesgue and having a finite integral of $\varphi^*\varphi$. If the integral of $\varphi^*\varphi$ is zero, $\varphi$ is called a *zero function*. Two functions $\varphi$ and $\psi$ will be considered as *equal*, if their difference $\varphi - \psi$ is a zero function. A *vector* or *element of Hilbert space* is defined as a class of functions $\varphi$ equal to a given function $\psi$. We shall not distinguish between the vector and the function $\psi$, because this logical distinction is physically unimportant.

The set of all vectors is called *Hilbert space*. It is a *complex vector space*, i.e. for every two elements $\varphi$ and $\psi$ a sum $\varphi + \psi$ is defined, and for any complex number $a$ a product $a\varphi$, having the usual properties. Moreover, for every pair $\varphi, \psi$ a *scalar product* is defined by

$$(2.1) \qquad \langle \varphi, \psi \rangle = \int \varphi^* \psi \, dq$$

integrated over $q$-space. Obviously, $\langle \varphi, \psi \rangle$ is complex conjugate to $\langle \psi, \varphi \rangle$, and for constant $a$ we have

$$(2.2) \qquad \langle \varphi, a\psi \rangle = a \langle \varphi, \psi \rangle,$$

$$(2.3) \qquad \langle a\varphi, \psi \rangle = a^* \langle \varphi, \psi \rangle,$$

$$(2.4) \qquad \langle \varphi, \psi_1 + \psi_2 \rangle = \langle \varphi, \psi_1 \rangle + \langle \varphi, \psi_2 \rangle,$$

$$(2.5) \qquad \langle \varphi_1 + \varphi_2, \psi \rangle = \langle \varphi_1, \psi \rangle + \langle \varphi_2, \psi \rangle.$$

Two functions $\varphi, \psi$ are called *orthogonal*, if their scalar product is zero. A special case of the scalar product is the *squared norm*

$$(2.6) \qquad \|\varphi\|^2 = \langle \varphi, \varphi \rangle = \int \varphi^* \varphi \, dq = \int |\varphi|^2 dq.$$

The squared norm of $\varphi + \psi$ is

$$\begin{aligned}
\|\varphi + \psi\|^2 &= \langle \varphi + \psi, \varphi + \psi \rangle \\
&= \langle \varphi, \varphi \rangle + \langle \varphi, \psi \rangle + \langle \psi, \varphi \rangle + \langle \psi, \psi \rangle.
\end{aligned}$$

If $\varphi$ and $\psi$ are orthogonal, this formula simplifies to the "Theorem of Pythagoras" (see Fig. 1)

$$(2.7) \qquad \|\varphi + \psi\|^2 = \|\varphi\|^2 + \|\psi\|^2$$

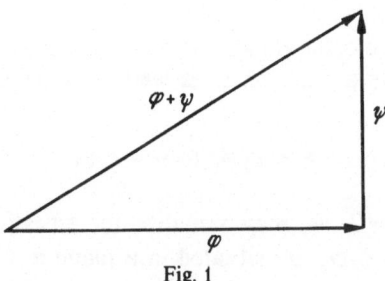

Fig. 1

which implies

(2.8) $$\|\varphi + \psi\|^2 \geqq \|\varphi\|^2 \quad \text{if} \quad \langle \varphi, \psi \rangle = 0.$$

*Orthogonalization.* If a sequence of functions $\psi_1, \psi_2, \ldots$ is not orthogonal, it may be replaced by an orthogonal sequence $\varphi_1, \varphi_2, \ldots$, such that all linear combinations $c_1 \psi_1 + \cdots + c_n \psi_n$ are also linear combinations of $\varphi_1, \ldots, \varphi_n$ and conversely. The definition of $\varphi_1, \varphi_2, \ldots$ is as follows:

$$\psi_1 = \varphi_1$$
$$\psi_2 = a\varphi_1 + \varphi_2$$
$$\psi_3 = b\varphi_1 + c\varphi_2 + \varphi_3, \quad \text{etc.}$$

If $\varphi_1$ happens to be zero, $\varphi_1$ may be skipped. If $\varphi_1$ is not zero, the coefficient $a$ may be determined in such a way that $\varphi_2$ becomes orthogonal to $\varphi_1$:

$$\langle \varphi_1, \varphi_2 \rangle = \langle \varphi_1, \psi_2 - a\varphi_1 \rangle = \langle \varphi_1, \psi_2 \rangle - a \langle \varphi_1, \varphi_1 \rangle = 0$$

which gives

$$a = \frac{\langle \varphi_1, \psi_2 \rangle}{\langle \varphi_1, \varphi_1 \rangle}.$$

Just so, if $\varphi_2$ is zero, it is skipped. If not, the coefficients $b$ and $c$ may be determined so as to make $\varphi_3$ orthogonal to $\varphi_1$ and $\varphi_2$:

$$b = \frac{\langle \varphi_1, \psi_3 \rangle}{\langle \varphi_1, \varphi_1 \rangle}, \quad c = \frac{\langle \varphi_2, \psi_3 \rangle}{\langle \varphi_2, \varphi_2 \rangle}$$

and so on.

In a subspace of finite dimension consisting of all sums $b_1 \psi_1 + \cdots + b_n \psi_n$, the orthogonalization process comes to an end after $n$ steps at most, and we obtain an *orthogonal basis* of the subspace.

Let $\varphi_1, \ldots, \varphi_n$ be orthogonal functions, all different from zero. Let us try to approximate a function $\psi$ as well as possible by a linear combination

(2.9)                          $$\psi' = c_1 \varphi_1 + \cdots + c_n \varphi_n.$$

In the case $n=2$, we may visualize the situation as follows. All vectors $\psi' = c_1 \varphi_1 + c_2 \varphi_2$ are situated in a plane $\pi$. Let $P$ and $P'$ be the end points of the vectors $\psi$ and $\psi'$. We have to find a point $P'$ in the plane as near as possible to $P$ (see Fig. 2).

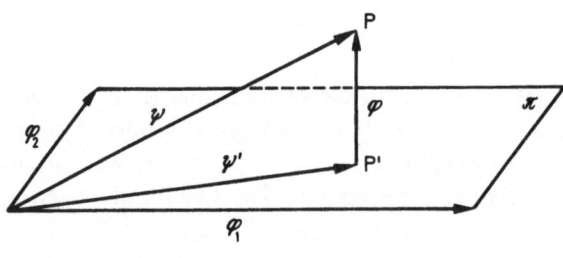

Fig. 2

The problem is solved by dropping a perpendicular $PP'$ from $P$ to the plane $\pi$. If $\psi'$ is the vector $OP'$ and $\varphi$ the vector $P'P$, we have

(2.10)                          $$\psi = \psi' + \varphi = \sum_1^n c_k \varphi_k + \varphi.$$

The problem is the same as that of orthogonalizing the sequence of vectors $\varphi_1, \ldots, \varphi_n, \psi$. The first $n$ vectors of the set are already orthogonal, and the last vector $\psi$ is replaced by $\varphi$, defined by (2.10). The coefficients $c_1, \ldots, c_n$ are determined in such a way that $\varphi$ is orthogonal to $\varphi_1, \ldots, \varphi_n$. This gives, as before,

(2.11)                          $$c_k = \frac{\langle \varphi_k, \psi \rangle}{\langle \varphi_k, \varphi_k \rangle}$$

or, if the vectors $\varphi_k$ are normed,

(2.12)                          $$c_k = \langle \varphi_k, \psi \rangle.$$

We still have to prove that $\psi'$ is the best approximation, or geometrically, that $P$ is nearer to $P'$ than to any other point $Q$ of the linear subspace spanned by $\varphi_1, \ldots, \varphi_n$. Let $\vec{OQ} = \psi''$ be any linear combination of $\varphi_1, \ldots, \varphi_n$. We have to prove:

$$\|\psi - \psi''\|^2 \geqq \|\psi - \psi'\|^2$$

or, substituting $\psi = \psi' + \varphi$,

$$\|\varphi + \psi' - \psi''\|^2 \geqq \|\varphi\|^2 .$$

Now this is an immediate consequence of (2.8), since $\psi' - \psi''$ is orthogonal to $\varphi$. Geometrically, $\vec{QP} = \psi - \psi''$ is the hypotenuse of a right-angled triangle $PP'Q$ with sides $\varphi$ and $\psi' - \psi''$ (Fig. 3).

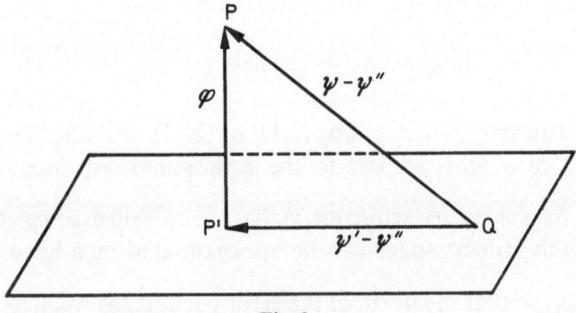

Fig. 3

The same inequality (2.8), applied to the sum (2.10), yields

(2.13) $$\|\psi\|^2 \geqq \|\psi'\|^2 .$$

Now the scalar product of $\psi' = c_1 \varphi_1 + \cdots + c_n \varphi_n$ by itself is, if the vectors $\varphi_k$ are normed,

(2.14) $$\| \psi' \|^2 = \langle \psi', \psi' \rangle = c_1^* c_1 + \cdots + c_n^* c_n = |c_1|^2 + \cdots + |c_n^2| .$$

From (2.13) and (2.14) we obtain the *Bessel inequality*

(2.15) $$|c_1|^2 + \cdots + |c_n|^2 \leqq \langle \psi, \psi \rangle .$$

In the special case $n = 1$, this inequality yields

$$|\langle \varphi_1, \psi \rangle|^2 \leqq \langle \psi, \psi \rangle$$

or, since $\varphi_1$ is normed,

$$|\langle \varphi_1, \psi \rangle|^2 \leqq \langle \varphi_1, \varphi_1 \rangle \langle \psi, \psi \rangle .$$

In this form, the inequality remains true if $\varphi_1$ is multiplied by an arbitrary constant $c$. Hence we have, for arbitrary functions $\varphi$ and $\psi$, the *Schwarz inequality*

(2.16)                         $$|\langle \varphi, \psi \rangle|^2 \leqq \langle \varphi, \varphi \rangle \langle \psi, \psi \rangle .$$

A consequence of this inequality is: If $\psi$ approximates $\psi_0$, so that the norm of $\psi - \psi_0$ becomes arbitrarily small, and if $\varphi$ is fixed, $\langle \varphi, \psi \rangle$ will approximate $\langle \varphi, \psi_0 \rangle$, i.e. the difference

$$\langle \varphi, \psi \rangle - \langle \varphi, \psi_0 \rangle = \langle \varphi, \psi - \psi_0 \rangle$$

becomes arbitrarily small. For, by (2.16), we have

$$|\langle \varphi, \psi - \psi_0 \rangle|^2 \leqq \|\varphi\|^2 \cdot \|\psi - \psi_0\|^2 .$$

The coefficients $c_k$, defined by (2.11) or (2.12), are called the *expansion coefficients* of $\psi$ with respect to the orthogonal sequence $(\varphi_1, \varphi_2, \ldots)$.

*Complete Sequences.* A sequence $\psi_1, \psi_2, \ldots$ is called *complete*, if every function $\psi$ in Hilbert space can be approximated by a linear combination $\sum_1^n a_k \psi_k$, so that the norm of the error becomes as small as we please:

(2.17)                         $$\left\| \psi - \sum_1^n a_k \psi_k \right\| < \varepsilon .$$

*If this is true, we may replace the sequence of $\psi_k$ by a normed orthogonal sequence $\varphi_1, \varphi_2, \ldots,$ and write*

(2.18)                         $$\left\| \psi - \sum_1^n b_k \varphi_k \right\| < \varepsilon .$$

Now replace $\Sigma b_k \varphi_k$ by the best possible approximation $\Sigma c_k \varphi_k$, in which the coefficients $c_k$ are given by (2.12). The approximation is at least as good, hence we have

(2.19)                         $$\left\| \psi - \sum_1^n c_k \varphi_k \right\| < \varepsilon .$$

If (2.19) holds, the series $\sum_1^\infty c_k \varphi_k$ is said to *converge in the mean* to the function $\psi$. Convergence in the mean is also called *strong convergence* in Hilbert space.

Substituting (2.10) into (2.19), we obtain $\|\varphi\| < \varepsilon$, hence

$$\|\varphi\| \to 0 \quad \text{for} \quad n \to \infty .$$

To (2.10) we may apply the theorem of Pythagoras:

$$(2.20) \qquad \|\psi\|^2 = \left\| \sum_1^n c_k \varphi_k \right\|^2 + \|\varphi\|^2 = \sum_1^n c_k^* c_k + \|\varphi\|^2 .$$

Letting $n$ go to infinity, we obtain the so-called *completeness relation*

$$(2.21) \qquad \langle \psi, \psi \rangle = \sum_1^\infty c_k^* c_k .$$

If this equation is satisfied for every function $\psi$, the sequence $\varphi_1, \varphi_2, \ldots$ is complete, and conversely.

Any complete set of orthogonal functions $\varphi_1, \varphi_2, \ldots$ is called an *orthogonal base* of the Hilbert space. The expansion coefficients $c_k = \langle \varphi_k, \psi \rangle$ are called the *coordinates* of the vector $\psi$ with respect to this base. They determine the vector $\psi$ uniquely, for if two vectors $\varphi$ and $\psi$ have the same coordinates, the difference $\varphi - \psi$ has coordinates zero, and hence, by (2.21), its norm is zero.

Now let $\varphi$ and $\psi$ have coordinates $b_k$ and $c_k$. We want to calculate the scalar product $\langle \varphi, \psi \rangle$. If we approximate $\psi$ by $\sum_1^n c_k \varphi_k$, the scalar product $\langle \varphi, \psi \rangle$ is approximated by

$$\langle \varphi, c_1 \varphi_1 + \cdots + c_n \varphi_n \rangle = c_1 \langle \varphi, \varphi_1 \rangle + \cdots + c_n \langle \varphi, \varphi_n \rangle$$
$$= c_1 b_1^* + \cdots + c_n b_n^* .$$

Hence, letting $n$ go to infinity, we obtain the formula

$$(2.22) \qquad \langle \varphi, \psi \rangle = \sum_1^\infty b_k^* c_k ,$$

a generalization of (2.21).

## § 3. Linear Operators

An operator $A$ transforming vectors into vectors is called a *linear operator* or *linear transformation*, if

$$(3.1) \qquad A(\varphi + \psi) = A\varphi + A\psi$$

and

(3.2) $$A(c\varphi) = cA\varphi$$

for constant $c$.

It is not required that $A\varphi$ is defined for all vectors $\varphi$. For instance, the momentum operators

(3.3) $$\frac{\hbar}{i}\frac{\partial}{\partial x}, \quad \frac{\hbar}{i}\frac{\partial}{\partial y}, \quad \frac{\hbar}{i}\frac{\partial}{\partial z}$$

are defined for differentiable functions only, and even if $\varphi$ is differentiable, $\partial\varphi/\partial x$ need not have a finite norm. However, it is required that $A\varphi$ is defined for a set of functions $\varphi$ which is dense in the Hilbert space, i.e. which approximates any function $\psi$ so that $\|\varphi - \psi\|$ can be made arbitrarily small.

Let $\varphi_1, \varphi_2, \ldots$ be orthogonal functions, to which the operator $A$ can be applied. The expansion coefficients of $A\varphi_k$:

(3.4) $$\langle \varphi_j, A\varphi_k \rangle = a_{jk}$$

will be called the *matrix elements* of the operator $A$ with respect to $\varphi_1, \varphi_2, \ldots$.

*Transition Probabilities.* For the interaction between an atom and the electromagnetic field, the *electrical momentum* of the atom is most important. The electrical momentum in the direction of the $x$-axis is

(3.5) $$X = \Sigma e_g x_g,$$

$x_g$ being the $x$-coordinate of the $g^{\text{th}}$ particle with respect to the centre of gravity of the atom. Similar expressions hold for the $y$- and $z$-components $Y$ and $Z$. To the classical quantity $X$ corresponds a quantum-mechanical operator $X$, viz. the multiplication of $\psi$ by $\Sigma e_g x_g$. The matrix elements of this operator with respect to the normed eigenfunctions of the Hamiltonian $H$ are

$$x_{jk} = \langle \varphi_j, X\varphi_k \rangle.$$

Let $E_j$ and $E_k$ be energy values belonging to the states $\varphi_j$ and $\varphi_k$. We may suppose $E_j < E_k$. If we restrict ourselves to dipole radiation, the probability for the atom to jump from the state $\varphi_k$ to the lower state $\varphi_j$ during the time $dt$, emitting a light quantum of frequency $\omega$, is[4]

$$(3.6) \qquad (|x_{jk}|^2 + |y_{jk}|^2 + |z_{jk}|^2) \frac{4\omega^3}{3\hbar c^3} \, dt \, .$$

The single matrix elements $x_{jk}, y_{jk}, z_{jk}$ determine the probabilities for emitting light quanta of given polarization. If there are several states $\varphi_j$ belonging to the same energy $E_j$, we have to form the sum of the expressions (3.6) for these states. The sum does not depend on the choice of the basis vectors $\varphi_j$ in the linear space of eigenvectors belonging to the energy $E_j$. The intensity of the emitted light is, of course, proportional to the transition probability.

If $x_{jk}, y_{jk}$ and $z_{jk}$ are zero for all pairs of vectors $\varphi_j$ and $\varphi_k$ belonging to the energies $E_j$ and $E_k$, the jump $E_k \rightarrow E_j$ does not occur: it is *forbidden*. This is not strictly true, for we have neglected multipole radiation.

*Symmetric Operators.* The operator $A$ is called *symmetric* (or *weakly self-adjoint*) if

$$(3.8) \qquad \langle \varphi, A\psi \rangle = \langle A\varphi, \psi \rangle$$

for any pair of functions to which $A$ can be applied. If $A$ is symmetric, the matrix $(a_{jk})$ has *Hermitean symmetry*, i.e.

$$(3.9) \qquad a_{jk}^* = a_{kj} \, .$$

Examples of symmetric operators are the operators $X$, $Y$, $Z$ just defined, and the mechanical momentum or impulse operators (3.3). The proof of (3.8) for the latter operators is given by partial integration. The boundary integrals occurring if the integration is extended to a finite domain vanish in the limit if the boundary goes to infinity. Similarly, the energy operator $H$ can be shown to be symmetric, provided it is applied to such functions $\varphi$ only, for which all derivatives $\varphi_x$, $\varphi_y$, $\varphi_z$ have finite integrals $\int \varphi_x^* \varphi_x dq$ etc.

We shall now prove two theorems on the eigenvalues and eigenfunctions of symmetric operators.

*1. The Eigenvalues of a Symmetric Operator are Real.*

*Proof.* Let $\psi$ be an eigenfunction of $A$:

$$A\psi = \lambda\psi \, .$$

---

[4] Formula (3.6) is implicit in Heisenberg's first paper on Quantum Mechanics in Z. Phys. **33** (1925). A justification of this formula by Quantum Field Theory was given by Jordan in his joint paper with Born and Heisenberg (Z. Phys. **35**). A still more satisfactory derivation of the same formula was given by Dirac in Proc. Roy. Soc. A 114 (1927).

By (3.8), we have

$$\langle \psi, A\psi \rangle = \langle A\psi, \psi \rangle \,,$$
$$\langle \psi, \lambda\psi \rangle = \langle \lambda\psi, \psi \rangle \,,$$
$$\lambda \langle \psi, \psi \rangle = \lambda^* \langle \psi, \psi \rangle \,.$$

Since $\langle \psi, \psi \rangle$ is not zero, $\lambda$ must be equal to $\lambda^*$, hence real.

2. *Eigenfunctions of a Symmetric Operator, Belonging to Different Eigenvalues, are Orthogonal.*

*Proof.* Let $\psi_1$ and $\psi_2$ be eigenfunctions of $A$, belonging to different eigenvalues:

$$A\psi_1 = \lambda_1 \psi_1 , \quad A\psi_2 = \lambda_2 \psi_2 , \quad \lambda_1 \neq \lambda_2 \,.$$

By (3.8), we have

$$\langle A\psi_1, \psi_2 \rangle = \langle \psi_1, A\psi_2 \rangle \,,$$
$$\langle \lambda_1 \psi_1, \psi_2 \rangle = \langle \psi_1, \lambda_2 \psi_2 \rangle \,.$$

Since $\lambda_1$ is real, we may write this as

$$\lambda_1 \langle \psi_1, \psi_2 \rangle = \lambda_2 \langle \psi_1, \psi_2 \rangle ,$$
$$(\lambda_1 - \lambda_2) \langle \psi_1, \psi_2 \rangle = 0 \,,$$
$$\langle \psi_1, \psi_2 \rangle = 0 \,.$$

## § 4. Hypermaximal Operators

The operators corresponding to measurable quantities in quantum mechanics are not only symmetric (which ensures that their eigenvalues are real), they are even *self-adjoint* or *hypermaximal* in the sense of von Neumann[5]. We shall explain what this means.

Let $A$ be any linear operator, operating on a set of functions $\psi$ dense in Hilbert space, and let $\varphi$ be a fixed vector. Suppose a vector $\chi$ exists such that, for all $\psi$ to which $A$ can be applied, the following equality holds:

(4.1) $$\langle \varphi, A\psi \rangle = \langle \chi, \psi \rangle \,.$$

If such a vector $\chi$ exists, it is uniquely determined by $\varphi$, and the mapping $\varphi \to \chi$ is a linear operator $A^*$, called the *adjoint* of $A$. Now if the adjoint

---

[5] See: J. von Neumann's fundamental paper: Allgemeine Eigenwerttheorie Hermetischer Funktionaloperatoren. Math. Ann. **102**, p. 49 (1929).

$A^*$ is exactly the same operator as $A$, which implies that it is defined for just the same functions $\psi$ and no more, the operator $A$ is called *self-adjoint* or *hypermaximal*.

If $A$ is hypermaximal, (4.1) implies

$$(4.2) \qquad\qquad \langle \varphi, A\psi \rangle = \langle A\varphi, \psi \rangle,$$

hence $A$ is symmetric. Conversely, if $A$ is symmetric, it very often can be extended so as to make the operator hypermaximal, but this extension is not always unique. Different extensions may give rise to different spectra. For instance, if $A$ is the operator $-d^2/dx^2$, applied to functions $\psi$ defined on the closed segment $(0, \pi)$ which are zero on the boundary together with their first derivatives, the operator is symmetric, but not hypermaximal: it has no eigenfunctions and no spectrum. If the domain of definition is extended to include all functions which are zero on the boundary, the eigenfunctions are $\sin nx$ and the eigenvalues $n^2$. If other boundary conditions are imposed, different eigenvalues may be obtained.

T. Kato [6] has proved that the Hamiltonian $H$ of an atom or molecule is self-adjoint.

J. von Neumann has proved that any self-adjoint operator has a *spectral resolution*. To explain what this means, let us first consider an operator $A$ having a pure point spectrum, e.g. the operator considered before:

$$A\psi = -\frac{d^2\psi}{dx^2} \quad (0 \le x \le \pi)$$

$$\psi(0) = \psi(\pi) = 0.$$

In this case, the normed and orthogonalized eigenfunctions $\varphi_1, \varphi_2, \ldots$ form a complete orthogonal sequence. Every vector $\psi$ can be expanded in a strongly convergent series

$$(4.3) \qquad\qquad \psi = \sum_1^\infty c_k \varphi_k, \quad c_k = \langle \varphi_k, \psi \rangle.$$

If $\psi$ belongs to the domain of the operator $A$, the vector $A\psi$ can also be expanded

$$A\psi = \Sigma b_k \varphi_k,$$

$$b_k = \langle \varphi_k, A\psi \rangle = \langle A\varphi_k, \psi \rangle$$

$$= \langle \lambda_k \varphi_k, \psi \rangle = \lambda_k \langle \varphi_k, \psi \rangle = \lambda_k c_k,$$

[6] T. Kato: Fundamental Properties of Hamilton operators of Schroedinger Type. Transactions Amer. Math. Soc. **70**, p. 195 (1951).

hence we have

(4.4) $$A\psi = \Sigma \lambda_k c_k \varphi_k.$$

Let $\lambda_1, \lambda_2, \ldots$ be the different eigenvalues. To every $\lambda$ we have a partial sum in (4.3)

$$\psi_\lambda = \Sigma c_k \varphi_k, \quad (\varphi_k \text{ belonging to } \lambda)$$

the summation extending over those $\varphi_k$ that belong to the eigenvalue $\lambda$. The eigenfunctions belonging to any $\lambda$ form a linear subspace $L_\lambda$ of the Hilbert space, and $\psi_\lambda$ is just the orthogonal projection of $\psi$ upon $L_\lambda$. Let $P_\lambda$ be the projection operator transforming $\psi$ into its projection $\psi_\lambda$. Now (4.3) may be written as

(4.5) $$\psi = \sum_\lambda \psi_\lambda = \sum_\lambda P_\lambda \psi.$$

Since this is true for every function $\psi$, the identity operator 1 is the sum of the projectors $P_\lambda$:

(4.6) $$1 = \sum_\lambda P_\lambda.$$

Similarly, (4.4) may be written as

(4.7) $$A = \sum_\lambda \lambda P_\lambda.$$

This is the "spectral resolution" of the operator $A$.

In order to show the usefulness of this spectral resolution, I shall give two applications.

1. *Functions of A.* Let $f(u)$ be any complex-valued function of the real variable $u$. Following von Neumann, we may define

(4.8) $$f(A) = \Sigma f(\lambda) P_\lambda.$$

The domain of definition of the operator $f(A)$ consists of those vectors $\psi$, for which the sum

$$\Sigma f(\lambda) P_\lambda \psi = \Sigma f(\lambda) \psi_\lambda$$

converges. For instance, if $f(u)$ is the function $f(u) = e^{-iut}$, the sequence

$$\sum_\lambda e^{-i\lambda t} \psi_\lambda$$

converges for every function $\psi$, hence the operator

$$e^{-iAt} = \Sigma e^{-i\lambda t} P_\lambda$$

is defined everywhere in Hilbert space. In fact, it is easy to see that the operator is unitary.

2. *Solving Schroedinger's Equation.* The Eq. (1.1) may be slightly simplified by putting $H = \hbar A$. It now reads

(4.9)
$$\frac{d\Psi}{dt} = -iA\Psi.$$

If $H$ has a pure point spectrum, so has $A$. If the eigenvalues of $H$ are $E = \hbar\omega$, those of $A$ are the frequencies and the spectral resolution of $A$ is $A = \sum_{\omega} \omega P_{\omega}$. Now the solution of (4.9) may be at once written down as

(4.10)
$$\Psi(t) = e^{-iAt}\Psi(0)$$

with

(4.11)
$$e^{-iAt} = \sum_{\omega} e^{-i\omega t} P_{\omega}.$$

In order to generalize the spectral resolution to continuous spectra, we have to replace the sums (4.6) and (4.7) by integrals. In general, there are no eigenfunctions belonging to single eigenvalues $\lambda$, but to any interval $\Delta\lambda$ defined by

$$\lambda_1 \leqq \lambda < \lambda_2$$

corresponds a linear subspace $L_{\Delta\lambda}$ of the Hilbert space. The orthogonal projection upon this subspace is a projection operator, which may be written as a difference of projection operators corresponding to intervals from $-\infty$ to $\lambda_1$ and $-\infty$ to $\lambda_2$:

$$\Delta P(\lambda) = P(\lambda_2) - P(\lambda_1).$$

Instead of the sums (4.6), (4.7), (4.8) and (4.11) we now have integrals

$$1 = \int_{-\infty}^{\infty} dP(\lambda)$$

$$A = \int_{-\infty}^{\infty} \lambda \, dP(\lambda)$$

$$f(A) = \int_{-\infty}^{\infty} f(\lambda) \, dP(\lambda)$$

$$e^{-iAt} = \int_{-\infty}^{\infty} e^{-i\lambda t} \, dP(\lambda),$$

and the solution of Schrödinger's equation can be written, as before, in the form

$$\Psi(t) = e^{-iAt} \Psi(0), \quad A = \hbar^{-1} H.$$

For a full explanation and proof of these formulae the reader may consult the book of M. H. Stone: Linear Transformations in Hilbert space (Amer. Math. Soc. Colloquium Publications **15**, Providence 1932).

## § 5. Separation of Variables

It often happens that the coordinates $q_1, \ldots, q_s$ of the particles under consideration may be separated into two sets $q_1, \ldots, q_n$ and $q_{n+1}, \ldots, q_s$, and that the energy operator (or an approximate, preliminary energy operator) $H$ consists of two parts:

$$H = H_1 + H_2,$$

where $H_1$ contains only functions of $q_1, \ldots, q_n$ and differentiations with respect to these variables, whereas $H_2$ contains only functions of and differentiations with respect to $q_{n+1}, \ldots, q_s$. For instance, $H_1$ and $H_2$ may be the energy operators of two atoms whose interaction is neglected. In these cases eigenfunctions of $H$ may be obtained as products

$$\varphi(q_1, \ldots, q_n)\, \psi(q_{n+1}, \ldots, q_s),$$

where $\varphi$ is an eigenfunction of $H_1$ and $\psi$ an eigenfunction of $H_2$. It is quite clear that these products are eigenfunctions of $H$, for we have

$$H \varphi\psi = (H_1 + H_2)\, \varphi\psi = (H_1 \varphi)\, \psi + \varphi(H_2 \psi)$$

$$= E_1 \varphi\psi + E_2 \varphi\psi = (E_1 + E_2)\, \varphi\psi.$$

Hence, the eigenvalue corresponding to $\varphi\psi$ is $E = E_1 + E_2$. The question is now, whether all eigenvalues of the system can be obtained in this way.

The answer is yes, if we suppose that the eigenfunctions $\varphi_1, \varphi_2, \ldots$ of $H_1$ form a complete orthogonal set.

*Proof.* Let $\chi$ be an eigenfunction of $H$. We may consider $\chi$ as a function of $q_1, \ldots, q_n$ and expand it in a series, convergent in the mean:

(5.1) $$\chi = \sum_{1}^{\infty} \varphi_v(q_1, \ldots, q_n)\, c_v(q_{n+1}, \ldots, q_s).$$

The coefficients $c_v$ are scalar products in the space of $q_1, ..., q_n$:

$$(5.2) \qquad\qquad c_v = \langle \varphi_v, \chi \rangle = \int \varphi_v^* \chi \, dV_n$$

with $dV_n = dq_1 ... dq_n$. Now how does $H_2$ operate on $c_v$? We suppose that the differentiations with respect to $q_{n+1}, ..., q_s$ occurring in $H_2$ can be effected under the sign of integration in (5.2). Thus we find

$$
\begin{aligned}
H_2 c_v &= \int \varphi_v^* (H_2 \chi) \, dV_n = \langle \varphi_v, H_2 \chi \rangle \\
&= \langle \varphi_v, H \chi \rangle - \langle \varphi_v, H_1 \chi \rangle \\
&= \langle \varphi_v, E \chi \rangle - \langle H_1 \varphi_v, \chi \rangle,
\end{aligned}
$$

because $H_1$ is a symmetric operator. Now $\varphi_v$ is an eigenfunction of $H_1$ with eigenvalue $E_v$, hence

$$
\begin{aligned}
(5.3) \qquad\qquad H_2 c_v &= E \langle \varphi_v, \chi \rangle - E_v \langle \varphi_v, \chi \rangle \\
&= (E - E_v) \langle \varphi_v, \chi \rangle \\
&= (E - E_v) c_v.
\end{aligned}
$$

It follows that $c_v$ is an eigenfunction of $H_2$ with eigenvalue

$$(5.4) \qquad\qquad E_v' = E - E_v.$$

Hence every eigenvalue $E$ is a sum $E_v + E_v'$, and the corresponding eigenfunctions $\chi$ are sums of products of eigenfunctions of $H_1$ and $H_2$, as in (5.1).

If we suppose that the eigenvalues $E_v$ corresponding to the eigenfunctions $\varphi_v$ go to infinity for $v \to \infty$, and that the $E_v'$ are bounded from below, it follows that any given $E$ can be decomposed into $E_v + E_v'$ only in a finite number of ways, and that the sums (5.1) contain only a finite number of terms.

If the spectrum of $H_1$ is not a pure point spectrum, our assumptions concerning the eigenfunctions $\varphi_1, \varphi_2, ...$ are not satisfied. Still, similar results may be obtained by a suitable modification of the method. We may e.g. replace the sums (5.1) by integrals as in § 4. The following example may serve to make this clear.

Three of the coordinates of an atom or molecule can always be separated from the others, viz. the coordinates of the center of gravity. To fix the ideas, consider an atom consisting of a nucleus and $f$ electrons. Let $q_0$ be the coordinates of the nucleus, and $q_1, ..., q_f$ the coordinates of the electrons. The eigenvalue problem is

$$(5.6) \qquad\qquad \left( \sum_0^f - \frac{\hbar^2}{2\mu_g} \Delta_g + U \right) \psi = E \psi.$$

We now introduce new coordinates, viz. the coordinates $q_c$ of the center of gravity, and the coordinates $q'_g$ of the electrons relative to this center:

$$M q_c = \mu_0 q_0 + \mu_1 q_1 + \cdots + \mu_f q_f$$

$$M = \mu_0 + \mu_1 + \cdots + \mu_f$$

$$q'_g = q_g - q_c \qquad\qquad (g = 1, 2, ..., f).$$

Equation (5.6) transforms into

$$(5.7) \quad \left\{ -\frac{\hbar^2}{2M} \Delta_c - \sum_1^f \frac{\hbar^2}{2\mu_g} \Delta'_g + \frac{\hbar^2}{2M} \left( \sum_1^f \frac{\partial}{\partial q'_g} \right)^2 + U \right\} \psi = E\psi .$$

Here $\frac{\hbar}{i} \Sigma \partial/\partial q'_g$ is the vector-operator corresponding to the sum of the momenta of the electrons with respect to the center of gravity, or, what amounts to the same, the momentum of the nucleus with respect to this center taken with the negative sign. The square in (5.7) is the scalar square of the vector.

If there are no external forces, the potential energy $U$ depends only on the relative coordinates $q'$. Hence, the variables $q'_1, ..., q'_f$ and $q_c$ can be separated, and eigenfunctions of the whole system may be obtained as products:

$$(5.8) \qquad\qquad \psi = \psi_1(q_c) \, \psi_2(q'_1, ..., q'_f).$$

For $\psi_1$ we obtain the Schrödinger equation of a free particle of mass $M$. The eigenfunctions are plane waves. Instead of a sum (5.1) we now have an integral as in (1.8). There are no eigenfunctions belonging to a single energy value $E$, but there are functions belonging to an arbitrarily small energy interval $\Delta E$, and they are obtained by integrating the products (5.8), in which the first factor $\psi_1$ is a plane wave, whereas $\psi_2$ has to satisfy the reduced Schrödinger equation

$$(5.9) \qquad \left\{ -\sum_1^f \frac{\hbar^2}{2\mu_g} \Delta_g + \frac{\hbar^2}{2M} \left( \sum_1^f \frac{\partial}{\partial q'_g} \right)^2 + U \right\} \psi = E_2 \psi .$$

For all problems concerning atomic spectra, it is sufficient to solve (5.9) and to leave aside the motion of the center of gravity.

In the simplest case $f = 1$, in which the atom has only one electron, Eq. (5.9) reduces to

$$(5.10) \qquad \left( -\frac{\hbar^2}{2\mu_1} \frac{\mu_0}{\mu_0 + \mu_1} \Delta + U \right) \psi = E_2 \psi .$$

This is just the Schrödinger equation for an electron attracted by a fixed nucleus, with a correction factor $\mu_0/(\mu_0 + \mu_1)$ to the mass. We now proceed to solve this equation.

## § 6. One Electron in a Central Field

In the Schrödinger equation of a single electron in a central field of force

$$(6.1) \qquad -\frac{\hbar^2}{2\mu} \Delta\psi - eV\psi = E\psi$$

the potential $V$ is a function of the radius $r$. In the case of a pure Coulomb field (Hydrogen atom or Helium ion) the function $V$ is

$$(6.2) \qquad V = \frac{Ze}{r}.$$

The Eq. (6.1) may be solved by the method of separation of variables. In polar coordinates $r, \Theta, \varphi$ the operator $\Delta$ can be written as

$$(6.3) \qquad \Delta = \frac{\partial^2}{\partial r^2} + \frac{2}{r}\frac{\partial}{\partial r} + \frac{1}{r^2} D,$$

$D$ being the differential operator

$$D = \frac{1}{\sin\Theta} \frac{\partial}{\partial\Theta} \sin\Theta \frac{\partial}{\partial\Theta} + \frac{1}{\sin^2\Theta} \frac{\partial^2}{\partial\varphi^2}.$$

Thus, eigenfunctions may be obtained as products

$$(6.4) \qquad \psi = f(r)\, Y(\Theta, \varphi).$$

The factors $f$ and $Y$ satisfy the differential equations

$$(6.5) \qquad DY = \lambda Y,$$

$$(6.6) \qquad -\frac{\hbar^2}{2}\left(\frac{\partial^2}{\partial r^2} + \frac{2}{r}\cdot\frac{\partial}{\partial r} + \frac{\lambda}{r^2}\right) f - eVf = Ef.$$

The eigenvalue problem (6.5) for the function $Y(\Theta, \varphi)$ does not depend on the form of the potential $V(r)$. It is well known that solutions of this equation can be obtained as spherical harmonics $Y_l^{(m)}$, and that the corresponding eigenvalue is

(6.7)                              $$\lambda = -l(l+1).$$

In order to prove, by the method of §5, that *all* eigenfunctions of (6.1) are sums of products (6.4), we first show that the eigenfunctions $Y_l^{(m)}$ of (6.5) form a complete orthogonal set of functions on the unit sphere.

The easiest way to define spherical harmonics $Y_l$ or $Y_l^{(m)}$ seems to be, to start with *forms* $U_l$, i.e. homogeneous polynomials of degree $l$ in $x, y, z$, satisfying the potential equation $\Delta U_l = 0$. These forms will be called *potential forms*. To obtain them, we introduce complex coordinates $x + iy$ and $x - iy$ instead of $x$ and $y$ and write

$$U_l = \Sigma \, c_{pq}(x + iy)^p (x - iy)^q z^{l-p-q}.$$

(summation over all pairs of non-negative integers $p, q$ with $p + q \leq l$). The condition $\Delta U_l = 0$ gives the recursion formula

$$4(p+1)(q+1)c_{p+1,q+1} + (l-p-q)(l-p-q-1)c_{pq} = 0.$$

Now let $U_l^{(m)}$ be the sum of those terms of $U_l$ for which the difference $p - q$ has a definite value $m$. The recursion formula determines the coefficients of $U_l^{(m)}$ but for a constant factor. The integer $m$ ranges from $-l$ to $+l$, hence there are just $2l+1$ linearly independent potential forms $U_l^{(m)}$. Expressing $U_l$ and $U_l^{(m)}$ in polar coordinates $r$, $\Theta$, $\varphi$, we obtain

(6.8)                         $$U_l = r^l Y_l, \qquad U_l^{(m)} = r^l Y_l^{(m)}$$

and

(6.9)                              $$Y_l^{(m)} = e^{im\varphi} f_l^{(m)}(\Theta).$$

The condition $\Delta U_l = 0$ yields, by (6.3)

(6.10)                             $$D Y_l = -l(l+1) Y_l,$$

hence the spherical harmonics $Y_l$ are eigenfunctions of (6.5) with eigenvalues $\lambda = -l(l+1)$.

The differential operator $D$ is a symmetric operator in the sense of §3, hence spherical harmonics $Y_l$ and $Y_{l'}$ of different degrees $(l \neq l')$ are orthogonal. Also, spherical harmonics $Y_l^{(m)}$ and $Y_l^{(m')}$ of the same degree but with $m \neq m'$ are orthogonal, because the factor

$$e^{-im\varphi}\, e^{im'\varphi} = e^{i(m'-m)}\, \varphi$$

yields zero upon integration with respect to $\varphi$.

We shall now prove: *Every continuous function on the unit sphere can be approximated uniformly, with an error $< \varepsilon$, by a sum of spherical harmonics $U_l$.* This implies the completeness of the set of functions $U_l^{(m)}$.

Let $f(\Theta, \varphi)$ be a continuous function. The function

$$v = r \cdot f(\Theta, \varphi)$$

is continuous in the whole space. In the closed cube

$$|x| \leqq 1, \quad |y| \leqq 1, \quad |z| \leqq 1$$

the function $v$ can be approximated, according to a theorem of Weierstrass, by a polynomial $P(x, y, z)$. This polynomial is a sum of forms. Now we shall prove that every form $F$ of degree $n$ can be expressed as a sum of potential forms multiplied by powers of $r$, as follows:

$$(6.11) \qquad F = U_n + r^2 U_{n-2} + r^4 U_{n-4} + \cdots + r^{2h} U_{n-2h}.$$

This is certainly true for forms of degree 0 or 1, since these are potential forms. Now suppose the assertion to be true for forms of degree $n - 2$; we shall prove it for forms of degree $n$. Since $\Delta F$ is a form of degree $n - 2$, we have

$$(6.12) \qquad \Delta F = U'_{n-2} + r^2 U'_{n-4} + \cdots .$$

Now put

$$(6.13) \qquad U'_l = (n - l)(n + l + 1) U_l \quad (l = n - 2, n - 4, \ldots)$$

and determine the form $U_n$ by (6.11). We have to prove that $U_n$ is a potential form. Operating with $\Delta$ on both sides of (6.11), we obtain

$$(6.14) \qquad \Delta F = \Delta U_n + \sum_{l=n-2k} (n - l)(n + l + 1)\, r^{2k-2} U_l$$

$$= \Delta U_n + U'_{n-2} + r^2 U'_{n-4} + \cdots .$$

From (6.12) and (6.14) we obtain $\Delta U_n = 0$, which proves our point. Hence the set of functions $U_l$ or $Y_l$ is complete on the unit sphere.

Now let $\psi$ be any eigenfunction of (6.1). Let $\psi$ be expanded in a series converging in the mean

(6.15) $$\psi = \sum_{0}^{\infty} c_{lm}(r)\, Y_l^{(m)}(\Theta, \varphi).$$

For the sake of convenience, suppose the $Y_l^{(m)}$ to be normed. Then the coefficients $c_{lm}(r)$ are scalar products $\langle Y_l^{(m)}, \psi \rangle$, calculated by integration over the unit sphere. One can prove, just as in § 5, that the functions $c_{lm}(r)$ are eigenfunctions of (6.6), and that the series (6.15) contains only a finite number of terms. Hence, every eigenfunction $\psi$ is a finite sum of products (6.4).

The number $l$ is called the *azimuthal quantum number*. The eigenvalue problem (6.6) depends on $\lambda$ and hence on $l$. For every value of $l$, the eigenvalues of (6.6) can be numbered, starting with the lowest one, by a *main quantum number* $n$ taking the values

$$n = l+1, l+2, \dots .$$

The *magnetic quantum number* $m$ takes, for every $n$ and $l$, the $2l+1$ values
$$m = l, l-1, \dots, -l.$$

The energy value $E$ does not depend on $m$, hence we have, for every energy value, $2l+1$ linearly independent eigenfunctions $f(r)\, Y_l^{(m)}$.

In the case of a Coulomb potential $V = Ze/r$ the eigenfunctions of (6.6) are confluent hypergeometric functions[7] and the eigenvalues depend only on $n$, not on $l$:

$$E_n = -\frac{B}{n^2}, \qquad B = \frac{Z^2 \mu e^4}{2\hbar^2}.$$

For $Z = 1$, this gives us the wellknown hydrogen terms

$$\frac{1}{\lambda} = \frac{E}{2\pi\hbar c} = \frac{R}{n^2}, \qquad R = \frac{e^4}{4\pi\hbar^3 c} = 109\,722 \text{ cm}^{-1}.$$

The transitions

$$E_n \to E_2, \quad E_n \to E_3, \quad E_n \to E_1$$

yield the Balmer-, Paschen- and Lymanseries of the hydrogen spectrum (see Fig. 4).

---

[7] See: e.g. Whittaker and Watson: Modern Analysis, Chapter 16: The Confluent Hypergeometric Function.

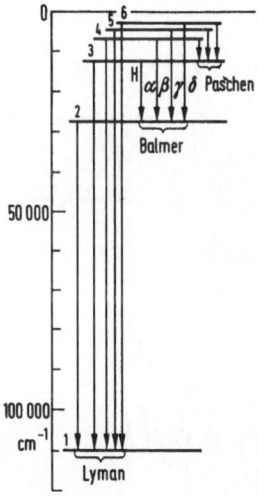

Fig. 4. Term Spectrum of the H-Atom

Next consider an atom such as Li, Na, K or Cs with one valence electron moving in a central field due to the nucleus and the other electrons. These electrons form a spherically symmetric cloud, which partly screens the field of the nucleus. In a second approximation, the polarization of the cloud by the valence electron may be taken into account. For very small values of $r$, the potential $V$ is nearly $Ze/r$: the cloud of electrons has no influence. For large values of $r$, the potential is approximately $e/r$: the nucleus and the cloud together act just like a hydrogen nucleus.

The resulting energy values are lower than the hydrogen eigenvalues, because the attraction is stronger than that of the hydrogen nucleus. The terms may be represented by a formula

(6.16)
$$\frac{1}{\lambda} = \frac{R}{(n-K)^2}$$

in which $K$ lies between 0 and 1/2 and does not depend very much on $n$. For $l = 0$ the terms lie considerably lower than the hydrogen terms, but for $l = 1, 2, \ldots$ the correction $K$ is nearly zero.

The formula (6.16) may be justified theoretically by perturbation theory, but originally it was found from experience. Spectroscopists found that the observed wave lengths of alkali lines may be obtained from terms like (6.16), arranged in series called $s$-, $p$-, $d$-, and $f$-series. These series correspond to the values 0, 1, 2, 3 of the quantum number $l$.

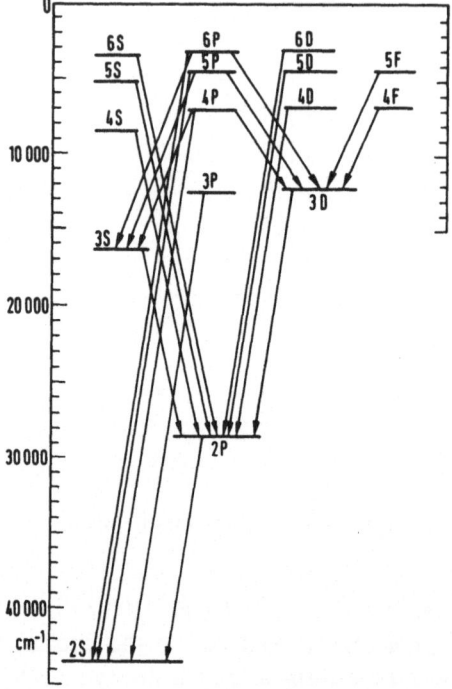

Fig. 5. Spectrum of Lithium

Since the main quantum number $n$ takes the values $l+1, l+2, ...$, we have, according to theory as well as to experience, the following terms:

$$l=0, \text{ } s\text{-series: terms } 1s, 2s, 3s, 4s, ...$$
$$l=1, \text{ } p\text{-series: terms } 2p, 3p, 4p, ...$$
$$l=2, \text{ } d\text{-series: terms } 3d, 4d, ...$$
$$l=3, \text{ } f\text{-series: terms } 4f, ...$$

Emission and absorption lines were always found to correspond to transitions from one series to a neighbouring series. Thus, $s$-terms combine with $p$-terms only, $p$-terms combine with $s$-terms and $d$-terms, etc. (see Fig. 5). In terms of the quantum number $l$, this rule can be formulated thus:

(6.17)                                    $l \rightarrow l \pm 1$.

A theoretical justification of this rule by means of group theory will be given in Chapter 3.

## § 7. Perturbation Theory

Consider an eigenvalue problem of the form

$$(7.1) \qquad (H_0 + \varepsilon V)\,\psi = E\psi$$

in which $\varepsilon$ is small. Suppose we know the eigenfunctions and eigenvalues of the undisturbed operator $H_0$. The problem is, to calculate the eigenfunctions and eigenvalues of the problem (7.1).

If we assume that the eigenvalues and eigenfunctions are power series in $\varepsilon$:

$$E = E_0 + \varepsilon E_1 + \cdots$$

$$\psi = \psi_0 + \varepsilon \psi_1 + \cdots$$

we may insert these into (7.1) and obtain conditions for the $E_k$ and $\psi_k$:

$$(7.2) \qquad (H_0 - E_0)\,\psi_0 = 0\,,$$

$$(7.3) \qquad (H_0 - E_0)\,\psi_1 = (E_1 - V)\,\psi_0\,,$$

$$(7.4) \qquad (H_0 - E_0)\,\psi_2 = (E_1 - V)\,\psi_1 + E_2\psi_0\,, \text{ etc.}$$

Condition (7.2) requires that $\psi_0$ be an eigenfunction of $H_0$ belonging to the eigenvalue $E_0$. This eigenfunction must be chosen in such a way that the conditions (7.3), etc. have solutions $\psi_1$, etc. Let $\varphi$ be any eigenfunction of $H_0$ belonging to the same eigenvalue $E_0$. Then the left-hand sides of (7.3), etc. are always orthogonal to $\varphi$:

$$\langle \varphi, (H_0 - E_0)\,\psi_1 \rangle = \langle (H_0 - E_0)\,\varphi, \psi_1 \rangle = 0\,,$$

hence the right-hand sides must also be orthogonal to $\varphi$. We thus obtain the conditions

$$(7.5) \qquad \langle \varphi, (E_1 - V)\,\psi_0 \rangle = 0 \quad \text{for all } \varphi, \text{ etc.}$$

Now suppose the eigenfunctions of $H_0$ with eigenvalue $E_0$ form a linear space $S_0$ spanned by normed orthogonal eigenvectors $\varphi_1, \varphi_2, \ldots, \varphi_n$. Then (7.5) can be written as

$$(7.6) \qquad \langle \varphi_\mu, (E_1 - V)\,\psi_0 \rangle = 0 \quad (\mu = 1, \ldots, n)\,.$$

This is a finite-dimensional eigenvalue problem for the unknown function

$$(7.7) \qquad \psi_0 = c_1 \varphi_1 + \cdots + c_n \varphi_n.$$

Putting $\langle \varphi_\mu, V \varphi_\nu \rangle = v_{\mu\nu}$ we may write (7.6) as

$$(7.8) \qquad E_1 c_\mu - \Sigma v_{\mu\nu} c_\nu = 0.$$

Hence $E_1$ is an eigenvalue of the finite matrix $(v_{\mu\nu})$, and $\psi_0$ as given by (7.7), is an eigenvector corresponding to this eigenvalue. The eigenvalues are solutions of the *secular equation*

$$(7.9) \qquad \begin{vmatrix} \lambda - v_{11} & v_{12} & \cdots & v_{1n} \\ v_{21} & \lambda - v_{22} & \cdots & v_{2n} \\ \cdots\cdots\cdots\cdots\cdots\cdots\cdots\cdots\cdots \\ v_{n1} & v_{n2} & \cdots & \lambda - v_{nn} \end{vmatrix} = 0.$$

The eigenvectors $\psi_0$ can be found by solving the linear equations (7.8), and $\psi_1$ can be calculated from the linear equation (7.3). The next eq. (7.4) can be treated by the same method. Thus, power series for $E$ and $\psi$ are obtained which formally satisfy condition (7.2). This is the *perturbation procedure* elaborated by Born[8] in 1925 and rediscovered by Schrödinger[9] in 1926. The treatment given here is from a book of Friedrichs[10].

The question now arises, whether the eigenvalues $E$ and eigenfunctions $\psi$ really are analytic functions of $\varepsilon$, which can be expanded in convergent power series. This question was first treated by F. Rellich in a fundamental series of papers in Math. Annalen **113, 116, 117,** and **118**. Assuming the operators $H_0$ and $\varepsilon V$ to satisfy certain regularity conditions, Rellich showed that the eigenvalues and eigenfunctions are in fact analytical functions of $\varepsilon$.

Rellich's investigations were continued from a more general point of view by Sz. Nagy, T. Kato and others. Friedrichs extended the theory to continuous spectra. For full references the reader may consult the book of T. Kato: Perturbation Theory for Linear Operators. Berlin-Heidelberg-New York: Springer 1966.

---

[8] Born, Heisenberg and Jordan: Z. Phys. **35**, p. 557, Chapter I, § 4.

[9] E. Schrödinger: Quantisierung als Eigenwertproblem III, Annalen der Physik **80**, p. 437.

[10] K. O. Friedrichs: Perturbation of Spectra in Hilbert Spaces. Amer. Math. Soc. (Providence 1965).

## § 8. Angular Momentum and Infinitesimal Rotations

In wave mechanics, the components of the moment of momentum of an atom or molecule are represented by the operators

$$\hbar L_x = \frac{\hbar}{i} \Sigma \left( y \frac{\partial}{\partial z} - z \frac{\partial}{\partial y} \right),$$

$$\hbar L_y = \frac{\hbar}{i} \Sigma \left( z \frac{\partial}{\partial x} - x \frac{\partial}{\partial z} \right),$$

$$\hbar L_z = \frac{\hbar}{i} \Sigma \left( x \frac{\partial}{\partial y} - y \frac{\partial}{\partial x} \right),$$

the summation extending over all particles. The square of the total moment of momentum is represented by the operator

$$\hbar^2 \mathscr{L}^2 = \hbar^2 (L_x^2 + L_y^2 + L_z^2).$$

The operators $L_x$, $L_y$, $L_z$ stand in a simple relation to spatial rotations. If the position of a single electron is subjected to a rotation about the $z$-axis with an infinitely small angle of rotation $d\alpha$, the changes of the coordinates $x$, $y$, $z$ are

$$dx = -y d\alpha, \quad dy = x d\alpha, \quad dz = 0.$$

The effect of a rotation $R$ upon a function $\psi(q) = \psi(x, y, z)$ is defined as follows: the function $\psi$ is transformed into a new function $\psi'$ such that

(8.2) $$\psi'(Rq) = \psi(q).$$

In other words, the value of the new function $\psi'$ at the new point $q' = Rq$ is defined as the value of the original function $\psi$ at the original point $q$. Instead of (8.2) we may also write

$$\psi'(q) = \psi(R^{-1}q),$$

$R^{-1}$ being the inverse rotation. Thus, in the case of an infinitely small rotation about the $z$-axis, we have

$$\psi'(x, y, z) = \psi(x - dx, y - dy, z - dz)$$

and hence

$$d\psi = \psi' - \psi = -\frac{\partial \psi}{\partial x} dx - \frac{\partial \psi}{\partial y} dy - \frac{\partial \psi}{\partial z} dz$$

$$= \left( y \frac{\partial \psi}{\partial x} - x \frac{\partial \psi}{\partial y} \right) d\alpha.$$

Hence, the increment $d\psi$ is found by applying to $\psi$ the operator

$$y \frac{\partial}{\partial x} - x \frac{\partial}{\partial y} = -\left( x \frac{\partial}{\partial y} - y \frac{\partial}{\partial x} \right)$$

and multiplying by the angle of rotation $d\alpha$. Just so, for a function $\psi(q_1, ..., q_f)$ of the coordinates of several particles, the increment $\delta\psi$ corresponding to an infinitely small simultaneous rotation of all particles about the Z-axis is found by applying to $\psi$ the operator

$$(8.3) \qquad I_z = -\Sigma \left( x \frac{\partial}{\partial y} - y \frac{\partial}{\partial x} \right) = -iL_z$$

(summation over the $f$ particles) and multiplying by $d\alpha$. For this reason, the operator (8.3) is called an *infinitesimal rotation* about the Z-axis. Just so, the operators

$$I_x = -iL_x, \qquad I_y = -iL_y$$

are called infinitesimal rotations about the X- and Y-axis.

The operators $I_x, I_y, I_z$ satisfy the commutation relations

$$(8.4) \qquad \begin{cases} I_y I_z - I_z I_y = I_x, \\ I_z I_x - I_x I_z = I_y, \\ I_x I_y - I_y I_x = I_z. \end{cases}$$

If the field of force is spherically symmetric, a rotation transforms eigenfunctions belonging to any energy level into eigenfunctions belonging to the same level. It follows that the infinitesimal rotations $I_x, I_y, I_z$ induce linear transformations in the linear space of eigenfunctions belonging to any energy level. In the case of a single electron in a spherically symmetric field the eigenfunctions are sums of products of functions $f(r)$ and spherical harmonics $Y_l^m$. The factor $f(r)$ is invariant with respect to rotations, so we need only investigate the linear transformation of the $Y_l^m$. Consider first the infinitesimal rotation $I_z$. Because $Y_l^m$ is equal to $e^{im\varphi}$ multiplied by a function of $\Theta$, we have

$$I_z Y_l^m = - im Y_l^m$$

hence

(8.5) $$L_z Y_l^m = m Y_l^m \,.$$

Thus, $Y_l^m$ belongs to the eigenvalue $m$ of $L_z$. We can also express this by saying: *In the state $f(r) \cdot Y_l^m$ the angular momentum $\hbar L_z$ has the value $\hbar m$.*

We now calculate the operator $\mathcal{L}^2$. In the case of one electron we have

$$- \mathcal{L}^2 = \left( y \frac{\partial}{\partial z} - z \frac{\partial}{\partial y} \right)^2 + \left( z \frac{\partial}{\partial x} - x \frac{\partial}{\partial z} \right)^2 + \left( x \frac{\partial}{\partial z} - z \frac{\partial}{\partial x} \right)^2$$

$$= r^2 \Delta - \left( x \frac{\partial}{\partial x} + y \frac{\partial}{\partial y} + z \frac{\partial}{\partial z} \right)^2 - \left( x \frac{\partial}{\partial x} + y \frac{\partial}{\partial y} + z \frac{\partial}{\partial z} \right) .$$

Introducing polar coordinates and applying (6.3), we obtain

(8.6) $$- \mathcal{L}^2 = D$$

and hence, because of (6.5) and (6.7)

(8.7) $$\mathcal{L}^2 f(r) Y_l^m = l(l+1) f(r) Y_l^m \,,$$

i.e.: *In the state $f(r) Y_l^m$ the operator $\mathcal{L}^2$ has the value $l(l+1)$.*

Physicists of the old school used to draw an angular momentum vector $\hbar \mathcal{L}$ of length $\hbar l$ in such a direction that its $z$-component has one of the possible values $\hbar m$ $(m = l, l-1, \ldots, -l)$.

*Normal Zeeman Effect.* If $\mathcal{H}$ is a constant magnetic field in the $z$-direction, a vector potential $\mathcal{A}$ may be defined by

$$A_x = \tfrac{1}{2} y H_z, \quad A_y = -\tfrac{1}{2} x H_z, \quad A_z = 0 .$$

The magnetic perturbation term in the Hamiltonian of an electron in this field is

$$W = \frac{e}{\mu c} (\mathcal{A} \cdot p)$$

with

$$p_x = \frac{\hbar}{i} \frac{\partial}{\partial x}, \text{ etc.}$$

Substituting, we obtain

(8.8) $$W = \kappa H_z L_z$$

with

$$\kappa = \frac{e\hbar}{2\mu c} = \text{Bohr's magneton}.$$

Let $H_0$ be the undisturbed central symmetric Hamiltonian of the electron. As we have seen, the eigenfunctions of $H_0$ can be chosen in such a way that they are at the same time eigenfunctions of $L_z$ with eigenvalue $m$, and hence eigenfunctions of the sum

$$H = H_0 + W = H_0 + \kappa H_z L_z$$

with eigenvalue

(8.9) $$E = E_0 + \kappa H_z m.$$

Hence the dislocation of the energy terms in the magnetic field is exactly $\kappa H_z m$.

It is easy to generalize this to an atom having several electrons. As we shall see in the next chapter, it is always possible to choose the eigenfunctions $\psi_m$ of the undisturbed Hamiltonian $H_0$ in such a way that a rotation $R_\alpha$ over an angle $\alpha$ about the $z$-axis transforms $\psi_m$ into $e^{-im\alpha}\psi_m$, where $m$ is an integer. Then we have, once more,

$$L_z \psi_m = m\psi_m,$$

and (8.9) follows as before.

The eigenvalue $m$ is called the *magnetic quantum number* of the atom, because the atom in a state $\psi_m$ behaves just like a magnet having a magnetic moment $m\kappa$ in the direction of the $z$-axis. The frequencies $\omega$ of the spectral lines corresponding to transitions $m \rightarrow m'$ are given by

(8.10) $$h\omega = E - E' = (E_0 - E_0') + \kappa H_z(m - m').$$

*The Selection Rule for m.* According to § 3, the transition probabilities and hence the intensities of the spectral lines are proportional to the squares of the absolute values of the matrix elements

$$\langle \psi_{m'}, X\psi_m \rangle, \quad \langle \psi_{m'}, Y\psi_m \rangle, \quad \langle \psi_{m'}, Z\psi_m \rangle$$

of the components $X, Y, Z$ of the electrical momentum of the atom. Instead of these, we may also consider the matrix elements of $X + iY$, $X - iY$ and $Z$, viz.

$$(X + iY)_{m'm} = \int \psi_{m'}^{*}(X + iY)\, \psi_m\, dq\,,$$

$$(X - iY)_{m'm} = \int \psi_{m'}^{*}(X - iY)\, \psi_m\, dq\,,$$

$$Z_{m'm} = \int \psi_{m'}^{*}\, Z \psi_m\, dq\,.$$

Consider e.g. the case of one electron. If the first integrand

$$\psi_{m'}^{*}(X + iY)\, \psi_m$$

is expressed in polar coordinates, it contains a factor

$$e^{i(-m'+1+m)\varphi}$$

which, integrated with respect to $\varphi$, gives zero unless $m' = m + 1$. Just so, the second integral is zero unless $m' = m - 1$, and the third unless $m' = m$. This gives us the selection rule:

(8.11)              $m' = m + 1,\quad m - 1\ \ \text{or}\ \ m\,.$

Moreover, in the case $m' = m$ the emitted light is polarized in the $z$-direction, whereas in the other two cases an observer in the $xy$-plane will observe linearly polarized light, and an observer in the $z$-direction circularly polarized light. The same thing holds in the case of several electrons. The difference $m - m'$, which enters into the expression $\hbar\omega$ according to (8.10), can only be 0 or $\pm 1$, hence there will be only 3 equidistant lines. This is the *Normal Zeeman Effect*. The anomalous Zeeman effect will be discussed, in connection with the "spinning electron", in Chapter 4.

Chapter II

# Groups and Their Representations

## § 9. Linear Transformations

Any set of objects (called vectors) $u$, $v$, ... that can be added and multiplied by real or complex numbers $b$, $c$, ... will be called a *real* or *complex vector space*, provided the usual rules of addition and multiplication are satisfied:

$$
\begin{array}{ll}
u + v = v + u & (u + v)c = uc + vc \\
u + (v + w) = (u + v) + w & u(b + c) = ub + uc \\
0 + u = u & u(bc) = (ub)c \\
-u + u = 0 & u1 = u .
\end{array}
$$

For instance, the Hilbert spaces defined in § 2 are complex vector spaces. The eigenfunctions belonging to a certain energy level always form a complex vector space.

A vector space $\mathscr{V}$ is called *n-dimensional*, if all vectors $v$ in $\mathscr{V}$ are linear combinations

$$v = e_1 c_1 + \cdots + e_n c_n$$

of $n$ linearly independent vectors $e_1, ..., e_n$. The *dimension n* does not depend on the choice of the basic vectors $e_1, ..., e_n$.

**Example.** The eigenfunctions of an atom or molecule belonging to an energy level $E$ form a finite-dimensional vector space.

In this chapter, all vector spaces will be complex, finite-dimensional vector spaces.

A *linear operator* or *linear transformation A* of a vector space $\mathscr{V}$ into itself or into another vector space is an operation transforming every vector $v$ into a vector $Av$ and having the following properties:

$$A(u + v) = Au + Av , \quad A(uc) = (Au)c .$$

Applying the transformation $A$ to the basic vectors one obtains

(9.1) $$Ae_k = \Sigma e_i a_{ik}.$$

As soon as the coefficients $a_{ik}$ are known, one can apply $A$ to any vector $v = \Sigma e_k c_k$, obtaining

$$Av = \Sigma\Sigma e_i a_{ik} c_k.$$

We may write this as $\Sigma e_i c_i'$, the coefficients $c_i'$ being

(9.2) $$c_i' = \Sigma a_{ik} c_k.$$

Hence, as soon as a definite basis $e_1, ..., e_n$ of the vector space is chosen, every linear transformation $A$ of $\mathscr{V}$ into itself is completely determined by its *matrix*

$$A = \begin{pmatrix} a_{11} & a_{12} & \cdots & a_{1n} \\ \vdots & & & \vdots \\ a_{n1} & a_{n2} & \cdots & a_{nn} \end{pmatrix}$$

Just so, a linear transformation $A$ of $\mathscr{V}$ into another vector space $\mathscr{W}$ is determined by a rectangular matrix $A$.

If two linear transformations $A$ and $B$ are applied one after the other: first $B$, next $A$, one obtains a product transformation $AB$:

$$(AB) v = A(Bv) = ABv.$$

The matrix of the product transformation $AB$ is obtained by applying it to the basic vectors $e_k$:

$$ABe_k = A(Be_k) = A(\Sigma e_j b_{jk})$$
$$= \Sigma(Ae_j)b_{jk} = \Sigma e_i a_{ij} b_{jk}.$$

The sign $\Sigma$ means: summation over every index that occurs twice. The matrix of the product transformation is the product $AB$ of the matrices $A$ and $B$. The matrix elements of $AB$ are

$$c_{ik} = \Sigma a_{ij} b_{jk}.$$

If the determinant of the matrix $A$ is not zero, the equations (9.2) can be solved for the $c_k$. The $c_k$ are now linear functions of the $c_j'$:

(9.3) $$c_k = \Sigma d_{kj} c_j'.$$

The formula (9.3) defines a linear transformation $D$, the *inverse* of $A$:

$$D = A^{-1}$$

such that $A^{-1}A$ transforms every vector $v$ into itself

$$A^{-1}Av = v.$$

Hence the product $A^{-1}A$, and just so $AA^{-1}$, is the *identical transformation* or *identity* $I$, which transforms every vector $v$ into itself. The matrix of $I$ is the *unit matrix*

$$I = \begin{pmatrix} 1 & 0 & \cdots & 0 \\ 0 & 1 & \cdots & 0 \\ \vdots & & & \vdots \\ 0 & 0 & \cdots & 1 \end{pmatrix}.$$

On the other hand, if the determinant Det $(A)$ is zero, the transformation $A$ has no inverse and is called *singular*.

If, instead of the $e_k$, new basic vectors $d_k$ are introduced by means of a non-singular linear transformation $P$ with matrix $P = (p_{jk})$:

$$d_k = Pe_k = \Sigma e_j p_{jk}$$

$$e_k = P^{-1}d_k = \Sigma d_h q_{hk}$$

the matrix of $A$ with respect to the new basis is calculated as follows

$$Ad_k = A\Sigma e_j p_{jk} = \Sigma e_i a_{ij} p_{jk}$$

$$= \Sigma d_h q_{hi} a_{ij} p_{jk},$$

hence the new matrix of $A$ is

(9.4)                                    $A' = P^{-1}AP.$

A *Hermitean form* on a vector space is an expression of the following kind

(9.5)                                    $H(v) = \Sigma h_{ik} c_i^* c_k$

with complex coefficients $h_{ik}$ satisfying the symmetry condition

$$h_{ki} = h_{ik}{}^*.$$

Because of this condition, the values $H(v)$ are real numbers. If $H(v)$ is positive for every vector $v \neq 0$, the form $H$ is called *positive*. An example is the *unit form*

$$H(v) = \Sigma c_k^* c_k .$$

Now let $H(v)$ be an arbitrary Hermitean form. Calculating $H(u + v)$, one obtains a sum of four terms:

$$H(u + v) = H(u) + H(v) + H(u, v) + H(v, u)$$

with

$$H(u, v) = \Sigma h_{ik} b_i^* c_k$$

and

$$H(v, u) = \Sigma h_{ik} c_i^* b_k .$$

The expression $H(u, v)$ is called the *scalar product* belonging to the form $H$. It has the following properties:

$$H(uc, v) = c^* H(u, v)$$
$$H(u, vc) = c H(u, v)$$
$$H(u + v, w) = H(u, w) + H(v, w)$$
$$H(u, v + w) = H(u, v) + H(u, w)$$
$$H(v, v) = H(v) ;$$
$$H(v, u) = H(u, v)^* .$$

If on a vector space a positive Hermitean form $H(v)$ is given, the space is called a *unitary vector space*. As long as only one positive Hermitean form is considered, one can drop the letter $H$ and write $\langle u, v \rangle$ and $\langle v, v \rangle$ instead of $H(u, v)$ and $H(v)$.

In § 2 an orthogonalization process was described for Hilbert spaces. The same process can also be applied in unitary vector spaces. Thus, one obtains the following

**Orthogonalization Theorem.** *If a positive Hermitean form* $\langle v, v \rangle$ *is given, one can always choose the basic vectors* $e_1, \ldots, e_n$ *in such a way that they are orthogonal*, i.e.

$$\langle e_i, e_k \rangle = 0 \quad for \quad i \neq k$$

*and normed*, i.e.

$$\langle e_k, e_k \rangle = 1 .$$

If the basic vectors are orthogonal and normed, the Hermitean form $H(v, v)$ becomes the unit form:

$$\langle v, v \rangle = \Sigma c_k^* c_k .$$

To every $m$-dimensional subspace $\mathscr{S}$ of a unitary vector space $\mathscr{V}$ a *totally orthogonal* subspace $\mathscr{S}^{\perp}$ exists, consisting of those vectors $w$ which are orthogonal to $v_1, \ldots, v_m$ and hence to all vectors $v$ of $\mathscr{S}$. According to the Orthogonalization Theorem, we may choose $v_1, \ldots, v_m$ as normed, orthogonal vectors, and we can find vectors $v_{m+1}, \ldots, v_n$ orthogonal to each other and to $v_1, \ldots, v_m$ such that $v_1, \ldots, v_m, \ldots, v_n$ form a basis of the whole space $\mathscr{V}$. If this is done, $v_{m+1}, \ldots, v_n$ form a basis of $\mathscr{S}^{\perp}$, hence the dimension of $\mathscr{S}^{\perp}$ is $n - m$. Every vector $u$ of $\mathscr{V}$ can be written, in only one way, as a sum $v + w$, where $v$ is in $\mathscr{S}$ and $w$ is in $\mathscr{S}^{\perp}$. This fact is expressed by saying: $\mathscr{V}$ is a *direct sum* of $\mathscr{S}$ and $\mathscr{S}^{\perp}$:

$$\mathscr{V} = \mathscr{S} \oplus \mathscr{S}^{\perp}.$$

A linear transformation $A$ is called *unitary*, if it leaves all scalar products unchanged

$$\langle Au, Av \rangle = \langle u, v \rangle.$$

This condition requires

$$\Sigma h_{ik} a_{ij}^{*} b_j^{*} a_{kl} c_l = \Sigma h_{jl} b_j^{*} c_l$$

identical in $b_j^{*}$ and $c_l$, hence

(9.6) $$\Sigma h_{ik} a_{ij}^{*} a_{kl} = h_{jl}.$$

If $A^{\dagger}$ denotes the transposed complex conjugate matrix of $A$, the condition (9.6) can be written as

$$A^{\dagger} H A = H$$

or, if $H$ is the unit matrix $I$

(9.7) $$A^{\dagger} A = I.$$

Equivalent to (9.7) is the condition that $A^{\dagger}$ be the reciprocal of $A$:

(9.8) $$A^{-1} = A^{\dagger}.$$

From (9.8) we see that a unitary transformation $A$ always has an inverse $A^{-1}$, hence any unitary transformation is non-singular.

A linear transformation $A$ is called *self-adjoint*, if it satisfies the condition

(9.9) $$\langle Au, v \rangle = \langle u, Av \rangle.$$

If the basic vectors are normed and orthogonal, the self-adjointness of $A$ can be expressed by the simple set of equations $a_{ik}^* = a_{ki}$ or

$$A^\dagger = A.$$

**Lemma.** *If a self-adjoint or unitary linear transformation $A$ transforms a linear subspace $\mathscr{S}$ of the vector space $V$ into itself, it also transforms the totally orthogonal subspace $\mathscr{S}^\perp$ into itself.*

*Proof.* First let $A$ be self-adjoint. If $v$ is an arbitrary vector of $\mathscr{S}$ and $w$ an arbitrary vector of $\mathscr{S}^\perp$, we have

$$\langle Aw, v \rangle = \langle w, Av \rangle = 0$$

hence $Aw$ is orthogonal to all vectors $v$ of $\mathscr{S}$, hence $Aw$ is in $\mathscr{S}^\perp$.

Next let $A$ be unitary. $A$ induces a unitary transformation of $\mathscr{S}$ into itself. Since a unitary transformation is non singular, the transformation maps $\mathscr{S}$ onto itself, i.e. all vectors of $\mathscr{S}$ can be written as $Av$ with $v$ in $\mathscr{S}$. Now we have for every $v$ in $\mathscr{S}$ and $w$ in $\mathscr{S}^\perp$

$$\langle Aw, Av \rangle = \langle w, v \rangle = 0$$

hence $Aw$ is orthogonal to all vectors $Av$ in $\mathscr{S}$, hence $Aw$ is in $\mathscr{S}^\perp$.

By means of this lemma we now prove:

**Theorem.** *Every self-adjoint or unitary transformation $A$ possesses a complete orthogonal set of $n$ eigenvectors $v_1, \ldots, v_n$.*

*Proof.* Any solution $\lambda$ of the secular equation

$$(9.10) \qquad \begin{vmatrix} a_{11} - \lambda & a_{12} & \cdots & a_{1n} \\ a_{21} & a_{22} - \lambda & \cdots & a_{2n} \\ \cdots & \cdots & \cdots & \cdots \end{vmatrix} = 0$$

gives rise to at least one eigenvector $v_1$. The coordinates $c_1, \ldots, c_n$ of $v_1$ are found by solving the linear set of equations

$$(a_{11} - \lambda) c_1 + \qquad a_{12} c_2 + \cdots = 0$$

$$a_{21} c_1 + (a_{22} - \lambda) c_2 + \cdots = 0$$

$$\vdots$$

$$a_{n1} c_1 + \qquad a_{n2} c_2 + \cdots = 0.$$

By our lemma, $A$ transforms the space $\mathscr{S}^{\perp}_{n-1}$ of all vectors orthogonal to $v_1$ into itself. Hence, by the same method, an eigenvector $v_2$ within $\mathscr{S}^{\perp}_{n-1}$ can be found, the secular equation being of degree $n-1$ only. Within $\mathscr{S}^{\perp}_{n-1}$, the vectors orthogonal to $v_2$ form a subspace $\mathscr{S}^{\perp}_{n-2}$, and so on.

If the orthogonal vectors $v_1, ..., v_n$ just found are used as basic vectors of $\mathscr{V}$, the matrix of the transformation $A$ becomes a diagonal matrix:

$$A = \begin{pmatrix} \lambda_1 & 0 & \cdots \\ 0 & \lambda_2 & \\ \vdots & & \ddots \end{pmatrix}.$$

Thus, the transformation $A$ is *diagonalized* or *transformed to principal axes*.

If $A$ is diagonalized, it is easy to see that every eigenvalue $\lambda$ of $A$ is one of the diagonal elements $\lambda_1, ..., \lambda_n$, and every eigenvector $v$ is a linear combination of eigenvectors $v_k$ belonging to eigenvalues $\lambda_k = \lambda$. In fact, suppose that $v$ is any eigenvector:

$$Av = v\lambda.$$

We can express $v$ by means of the $v_k$:

$$v = v_1\alpha_1 + \cdots + v_n\alpha_n.$$

Applying $A$ to this sum, one obtains

$$v_1\lambda_1\alpha_1 + \cdots + v_n\lambda_n\alpha_n = (v_1\alpha_1 + \cdots + v_n\alpha_n)\lambda.$$

Equating coefficients, one gets

$$\alpha_k\lambda_k = \alpha_k\lambda \quad (k = 1, ..., n).$$

If $\alpha_k$ is not zero, this implies

$$\lambda_k = \lambda.$$

Hence all $\alpha_k$ are zero except those for which $\lambda_k$ is equal to $\lambda$, and $v$ is a linear combination of basic vectors $v_k$ with $\lambda_k = \lambda$.

If $A$ is self-adjoint, all eigenvalues $\lambda_k$ are real numbers. On the other hand, if $A$ is unitary, the absolute values of the $\lambda_k$ are equal to one, hence we may write

$$\lambda_k = e^{i\beta_k}.$$

**Theorem.** *Any set of commuting self-adjoint or unitary linear transformations can be simultaneously diagonalized.*

*Proof.* If all matrices belonging to transformations of the set are multiples $\lambda I$ of the unit matrix, the assertion is trivial. Hence, let $A$ be a transformation of the set whose matrix is not diagonal. Let

$$v_1, v_2, \ldots, v_h; \ w_1, w_2, \ldots, w_k; \ \ldots$$

be a complete set of orthogonal eigenvectors $A$, with eigenvalues

$$\lambda_1, \lambda_1, \ldots, \lambda_1; \lambda_2, \lambda_2, \ldots \lambda_2; \ \ldots \ (\lambda_1 \neq \lambda_2, \text{etc.})\,.$$

To the eigenvalue $\lambda_1$ belongs a linear subspace $\mathscr{S}_h = (v_1, \ldots, v_h)$, to $\lambda_2$ a subspace $\mathscr{S}_k = (w_1, \ldots, w_k)$, etc. Now if another transformation $B$ commutes with $A$, it must transform $\mathscr{S}_h$ into itself, $\mathscr{S}_k$ into itself, etc.; for if $v$ is a vector in $\mathscr{S}_h$ we have

$$AB\,v = BA\,v = Bv\,\lambda_1$$

hence $Bv$ is an eigenvector with eigenvalue $\lambda_1$, i.e. $Bv$ is in $\mathscr{S}_h$.—Now supposing the theorem to be true for spaces $\mathscr{V}'$ of smaller dimension than the given vector space $\mathscr{V}$, it follows that it is true for $\mathscr{S}_h$ and for $\mathscr{S}_k$, etc., hence our set of transformations can be diagonalized in $\mathscr{S}_h$ and in $\mathscr{S}_k$, etc. Thus our theorem follows.

The left-hand side of the secular equation (9.10) is the determinant of the matrix $A - \lambda I$. If we multiply it by $(-1)^n$, we obtain the determinant of $\lambda I - A$:

$$\text{Det}\,(\lambda I - A) = (-1)^n \,\text{Det}\,(A - \lambda I)\,.$$

This determinant is a polynomial in $\lambda$. The coefficient of the highest power $\lambda^n$ is 1, that of $-\lambda^{n-1}$ is the *trace* of $A$:

$$\text{Tr}\,(A) = a_{11} + \cdots + a_{nn}\,,$$

whereas the constant term is

$$\text{Det}\,(-A) = (-1)^n \,\text{Det}\,(A)\,.$$

The polynomial $\text{Det}\,(\lambda I - A)$ is called the *characteristic polynomial* of $A$. It is invariant with respect to any change of basis; this is seen as follows:

$$\begin{aligned}
\text{Det}\,(\lambda I - P^{-1}AP) &= \text{Det}\,P^{-1}(\lambda I - A)P \\
&= \text{Det}\,(P^{-1}) \cdot \text{Det}\,(\lambda I - A) \cdot \text{Det}\,(P) \\
&= \text{Det}\,(\lambda I - A)\,.
\end{aligned}$$

Hence all coefficients of the characteristic polynomial are invariants. In particular we have

$$\mathrm{Tr}(P^{-1}AP) = \mathrm{Tr}(A)$$

and

$$\mathrm{Det}(P^{-1}AP) = \mathrm{Det}(A).$$

## § 10. Groups

A set $\mathscr{G}$ of elements $a, b, \ldots$ is called a *group*, if the following four conditions are satisfied:

1. For every pair of elements $a$, $b$ a *product* $a \cdot b = ab$ is defined as an element of $\mathscr{G}$.

2. Associative law: $ab \cdot c = a \cdot bc$.

3. A *unity* $e$ (or 1) is defined within $\mathscr{G}$, such that for every $a$ in $\mathscr{G}$

$$ae = ea = a.$$

4. To every $a$ there is an element $a^{-1}$ in $\mathscr{G}$ such that

$$aa^{-1} = a^{-1}a = e.$$

The group is called *Abelian* if all its elements commute:

$$ab = ba \quad \text{for every } a \text{ and } b.$$

If the elements of the group are linear transformations or permutations, and if the product $a \cdot b$ is defined as the transformation or permutation obtained by applying first $b$ and next $a$, the associative law is automatically fulfilled. In order to fulfill 3. and 4., we have to define $e$ as the identity and $a^{-1}$ as the inverse transformation to $a$. Hence we have:

*A non empty set of linear transformations or permutations is a group, if it contains together with any two elements a and b their product ab and with every element a its inverse $a^{-1}$.*

**Examples.** 1) The rotations of the real 3-dimensional space leaving fixed the origin $O$ form a non-abelian group: the *rotation group* $\mathcal{O}_3$ or $\mathrm{SO}(3, \mathbb{R})$. The letters have the following meaning:

$$\mathrm{SO} = \text{Special Orthogonal},$$
$$\mathbb{R} = \text{Field of Real Numbers}.$$

2) The rotations about a fixed axis form an Abelian group, which may be denoted by $\mathcal{O}_2$, or SO$(2, \mathbb{R})$.

3) The *Lorentz group* consists of those non-singular transformations of a real 4-dimensional vector space, which leave the quadratic form

$$x^2 + y^2 + z^2 - c^2 t^2$$

unchanged and which do not interchange past and future.

4) the *special linear groups* SL$(\mathbb{R}, n)$ and SL$(\mathbb{C}, n)$ are the groups of all linear transformations of determinant one, with real or complex coefficients ($\mathbb{R}$ = real field, $\mathbb{C}$ = Complex Field).

5) The *special unitary group* SU$(n)$ is the group of unitary transformations of a complex $n$-dimensional vector space with determinant one.

6) The *symmetric group* $\mathcal{S}_n$ consists of all one-to-one permutations of $n$ objects.

For these permutations the following notation is useful. The permutation of the numbers 1 2 3 4 5 which transforms 1 into 2 and 2 into 3 and 3 into 1 and which interchanges 4 and 5 is denoted by

(10.1)                        $P = (123)\,(45)$.

This notation is particularly convenient if one wants to calculate the permutations $QPQ^{-1}$. For instance, if $Q$ is the cyclic permutation (345), and if $P = (123)\,(45)$ as before, all one has to do to obtain $QPQ^{-1}$ is to permute the numbers 3 4 5 cyclically in the formula for $P$:

$$QPQ^{-1} = (124)\,(53)\,.$$

If $s$ is a fixed element of a group $G$, the elements $tst^{-1}$ are called *conjugate* to $s$. They form a *class of conjugate group elements*. For instance, the permutations conjugate to (123) (45) are all products of the form

$$(ijk)\,(lm)\,.$$

Every permutation can be written as a product of *transpositions* $(ik)$, which interchange only two objects $i$ and $k$. For instance, the cyclic permutation (1234) can be written as

$$(1234) = (12)\,(23)\,(34)\,.$$

The *even permutations*, which can be written as products of an even number of transpositions, form a group, the *alternating group* $\mathscr{A}_n$.

In Abelian groups, the notation $a \cdot b$ is often replaced by $a + b$. In such an *additive group* one writes 0 instead of 1, and $-a$ instead of $a^{-1}$. Thus, the conditions for an additive groupe are:

1') $\quad a + b = b + a$ is an element of $\mathscr{G}$,

2') $(a + b) + c = a + (b + c)$,

3') $\quad 0 + a = a$,

4') $\quad -a + a = 0$.

**Example.** Every vector space is an additive group.

In a vector space we have two operations: addition and multiplication by (real or complex) numbers $\vartheta$. Quite generally, suppose we have an additive group $\mathscr{G}$ and a set of "multiplicators" or "operators" $\vartheta$ such that

5') For every operator $\vartheta$ and every group element $a$, the "product" $\vartheta a$ is a definite element of $G$,

6') $\vartheta(a + b) = \vartheta a + \vartheta b$.

Under these conditions, $\mathscr{G}$ is called a "group with operators".

If a non-empty subset $\mathscr{H}$ of a group $\mathscr{G}$ is again a group, it is called a subgroup of $\mathscr{G}$. In order to be a subgroup of $\mathscr{G}$, the subset $\mathscr{H}$ must satisfy two conditions:

a) if $a$ and $b$ are in $\mathscr{H}$, $ab$ is in $\mathscr{H}$;

b) if $a$ is in $\mathscr{H}$, $a^{-1}$ must be in $\mathscr{H}$.

If $\mathscr{G}$ is written as an additive group, any subgroup $\mathscr{H}$ is required to contain $a + b$ and $-a$.

In the case of a group with operators only *admissible subgroups* are considered, which contain, together with any element $a$, all products $\vartheta a$. *Example*: if a vector space is considered as an additive group having the real or complex numbers as operators, the admissible subgroups are just the linear subspaces.

The *center* of a group $\mathscr{G}$ is the subgroup consisting of those elements $z$ which commute with all group elements $a$:

$$za = az \quad \text{for all } a \text{ in } \mathscr{G}.$$

The *residue classes* $b\mathscr{H}$ of a subgroup $\mathscr{H}$ in $\mathscr{G}$ are obtained by multiplying one element $b$ of $\mathscr{G}$ by all elements of $\mathscr{H}$. Two elements $b$ and $c$ belong to the same residue class if and only if $b^{-1}c$ is in $\mathscr{H}$. Two different residue classes have no element in common, and the union of all residue classes is the whole group $\mathscr{G}$. *Example*: The alternating group $\mathscr{A}_n$ has just two residue classes in $\mathscr{S}_n$, viz. the classes of even and of odd permutations.

If the elements of one residue class $a\mathscr{H}$ are multiplied by the elements of another $b\mathscr{H}$, the result is not always a residue class. To make sure that $a\mathscr{H} \cdot b\mathscr{H}$ is always a residue class, we must suppose that $\mathscr{H}$ is identical with all its conjugate subgroups $a\mathscr{H}a^{-1}$:

$$a\mathscr{H}a^{-1} = \mathscr{H}, \quad \text{or} \quad a\mathscr{H} = \mathscr{H}a \text{ for all } a \text{ in } \mathscr{G}.$$

Such subgroups $\mathscr{H}$ are called *normal divisors* of $\mathscr{G}$.

**Example.** $\mathscr{A}_n$ is a normal divisor of $\mathscr{S}_n$.

If $\mathscr{H}$ is a normal divisor of $\mathscr{G}$, the product $a\mathscr{H} \cdot b\mathscr{H}$ is the residue class $ab\mathscr{H}$, for we have

$$a\mathscr{H} \cdot b\mathscr{H} = a(\mathscr{H}b)\mathscr{H} = a(b\mathscr{H})\mathscr{H} = ab\mathscr{H}^2 = ab\mathscr{H}.$$

Thus, the residue classes form a group: the *factor group* or *residue class group* $\mathscr{G}/\mathscr{H}$.

In an Abelian group, e.g. in a vector space, every subgroup $\mathscr{H}$ is a normal divisor, hence the factor group $\mathscr{G}/\mathscr{H}$ can always be formed. If $\mathscr{G}$ is an additive group with operators and if $\mathscr{H}$ is an admissible subgroup, the residue classes $a + \mathscr{H}$ can be multiplied by the operators $\vartheta$, hence the residue class group $\mathscr{G}/\mathscr{H}$ is again a group with operators.

**Example.** Let $\mathscr{G}$ be a vector space $(e_1, ..., e_n)$, and let $\mathscr{H}$ be a subspace $(v_1, ..., v_m)$. The basis $v_1, ..., v_m$ of the subspace can be extended to a basis $(v_1, ..., v_n)$ of the whole space by adjoining $n - m$ new vectors $v_{m+1}, ..., v_n$. In order to obtain a residue class we have to add to one vector

$$v = v_1 c_1 + \cdots + v_m c_m + \cdots + v_n c_n$$

all vectors $v_1 b_1 + \cdots + v_m b_m$ of the subspace. The resulting residue class consists of all vectors

$$v = v_1 a_1 + \cdots + v_m a_m + v_{m+1} c_{m+1} + \cdots + v_n c_n$$

with variable $a_1, ..., a_m$, but fixed $c_{m+1}, ..., c_n$. As a representant of the residue class we may take the vector

$$v_{m+1} c_{m+1} + \cdots + v_n c_n.$$

Hence, the residue class group is a vector space of $n - m$ dimensions.

A *homomorphism* is a mapping $a \to a'$ of a group $\mathscr{G}$ into another group $\mathscr{G}'$ mapping products upon products: $ab \to a'b'$.

If the maps $a'$ of the elements of $\mathscr{G}$ cover the whole group $\mathscr{G}'$, the homomorphism is called a homomorphism of $\mathscr{G}$ *on to* $\mathscr{G}'$, and $\mathscr{G}'$ is said to be a *homomorphic map* of $\mathscr{G}$.

**Example.** Let $\mathscr{G}$ be the symmetric group $\mathscr{S}_n$. If all even permutations are mapped upon the number $+1$ and all odd permutations upon $-1$, we obtain a homomorphism of $\mathscr{S}_n$ on to the group consisting of the elements $+1$ and $-1$.

If different elements of $\mathscr{G}$ are always mapped upon different elements of $\mathscr{G}'$, the homomorphism is called an *isomorphism*. If the mapping is an isomorphism of $\mathscr{G}$ on to $\mathscr{G}'$, the groups $\mathscr{G}$ and $\mathscr{G}'$ are called isomorphic and we write

$$\mathscr{G} \cong \mathscr{G}' .$$

Now let $\mathscr{G}$ and $\mathscr{G}'$ be additive groups with operators, both having one and the same set of operators $\vartheta$. A homomorphism of $\mathscr{G}$ into $\mathscr{G}'$ maps $a + b$ upon $a' + b'$. An *operator-homomorphism* is a homomorphism that maps $\vartheta a$ upon $\vartheta a'$. The terms *operator-isomorphism* and *operator-isomorphic* are now clear. The sign $\cong$ is used also for operator-isomorphy.

**Examples.** 1) Every linear transformation of a real or complex vector space is an operator homomorphism, as long as only real or complex numbers are taken as operators $\vartheta$.

2) Any two vector spaces of the same dimension over the same field are operator-isomorphic.

If a homomorphism of $\mathscr{G}$ on to $\mathscr{G}'$ is given, the elements of $\mathscr{G}$ which are mapped upon 1 form a normal divisor $\mathscr{N}$ of $\mathscr{G}$, and the elements which have a given map $a'$ always form a residue class $a\mathscr{N}$. Thus, we have a one-to-one correspondence between residue classes $a\mathscr{N}$ and elements $a'$ of $\mathscr{G}'$. This correspondence is an isomorphism. Hence we have:

**Homomorphy Theorem.** *If $\mathscr{G}'$ is a homomorphic map of $\mathscr{G}$, then $\mathscr{G}'$ is isomorphic to a factor group $\mathscr{G}/\mathscr{N}$, where $\mathscr{N}$ is the normal divisor consisting of the elements of $\mathscr{G}$ whose map is* 1.

In the example of the symmetric group $\mathscr{G} = \mathscr{S}_n$ given before, the normal divisor $\mathscr{N}$ is the alternating group $\mathscr{A}_n$. It follows that the factor group $\mathscr{S}_n/\mathscr{A}_n$ is isomorphic to the group $\mathscr{G}'$ consisting of $+1$ and $-1$.

The Homomorphy Theorem holds without any change for operator-homomorphisms.

An important class of homomorphisms is formed by the re-presentations of groups. Let $\mathcal{G}$ be any group and let the elements $a$ of $\mathcal{G}$ be mapped upon linear transformations $A$ of a vector space $\mathcal{V}$ or, what amounts to the same thing, upon matrices $A$, such that the map of $ab$ is $AB$. Such homomorphisms are called *representations $\varrho$ of $\mathcal{G}$ by linear transformations or by matrices*. The dimension $n$ of the space $\mathcal{V}$ is called the *degree* of the representation $\varrho$. If the mapping $\varrho(a \rightarrow A)$ is one-to-one, we have a *true representation*. Because of the Homomorphism Theorem, any representation $\varrho$ of $\mathcal{G}$ is a true representation of a certain factor group $\mathcal{G}/\mathcal{N}$.

The application of the theory of Group Representations to Quantum Mechanics is based upon the following idea.

Schrödinger's differential equation

$$H\psi = E\psi$$

is invariant with respect to certain transformations of the variables $x, y, z, \ldots$, such as

1. permutations of the coordinates of electrons and of equal nuclei,

2. translations, rotations and reflections, which leave unchanged the field of force and hence the Hamiltonian $H$. Let us consider a few special cases of 2:

2a. If we have an atom, in which the nucleus is regarded as fixed, we have to consider the reflections and rotations leaving invariant this point.

2b. If the atom is placed in a homogeneous electric or magnetic field, the group of rotations about a fixed point must be replaced by the group of rotations about a fixed axis. There are also reflections which leave the field invariant.

2c. If we have a two-atom molecule, and if we consider (in a first approximation) the nuclei as being fixed in space, the rotations to be considered are rotations about the line connecting the nuclei, and the reflections are with respect to the planes passing through this line. If the two nuclei have equal atomic numbers and hence equal charges, we may also consider the reflection with respect to a plane perpendicular to the axis, which interchanges the two nuclei.

The permutations, rotations etc. just considered, which leave invariant the Hamiltonian $H$, always form a group $\mathcal{G}$. For: if the transformations $A$ and $B$ leave $H$ invariant, so does $AB$, and so does $A^{-1}$. The identity $I$ trivially leaves $H$ invariant.

The transformations of this group can be applied to the wave function $\psi$, as follows. Let $q_1, \ldots, q_f$ be the coordinates of the $f$ electrons of the atom or molecule under consideration, the single letter $q$ standing for a triple of three coordinates $x$, $y$, $z$. If a spatial transformation $T$ transforms the points $q_1, \ldots, q_f$ into points $q'_1, \ldots, q'_f$, the transformed wave function $\psi' = T\psi$ will be defined (as in § 8) by the formula

$$\psi'(q'_1, \ldots, q'_f) = \psi(q_1, \ldots, q_f)$$

or, what amounts to the same thing, by

$$\psi'(q_1, \ldots, q_f) = \psi(T^{-1}q_1, \ldots, T^{-1}q_f).$$

It is easy to see that this transformation of functions is a linear transformation:

$$(\varphi + \psi)' = \varphi' + \psi',$$

$$(c\varphi)' = c\varphi'.$$

If two transformations $S$ and $T$ are applied one after the other (first $T$, next $S$), we have

$$(ST)\psi = S(T\psi).$$

Since these transformations leave invariant Schrödinger's differential equation, they transform solutions $\psi$ into solutions $\psi'$ belonging to the same energy value. Hence:

*The eigenfunctions of any given energy level are linearly transformed by the group $\mathscr{G}$, and these transformations form a representation of the group $\mathscr{G}$.*

If we succeed in classifying all possible representations of the groups in question, this will give us at the same time a classification of the eigenfunctions and energy levels of the atoms and molecules. This classification forms, as we shall see, the theoretical basis of spectroscopy.

## § 11. Equivalence and Reducibility of Representations

If a representation $\varrho$ of a group $\mathscr{G}$ by linear transformations of a vector space $\mathscr{V}$ is given, it is often useful to consider $\mathscr{V}$ as an additive group with operators. The operators are the complex numbers α, β, ... and the group elements $a$, $b$,.... If the group element $a$ is represented by a linear transformation $A$, the product $av$ is defined as

$$av = Av$$

The fact that $a \to A$ is a representation of $\mathscr{G}$, i.e. that to $ab$ corresponds $AB$, is expressed by the formula

$$(ab)v = a(bv).$$

The notation $av$ is very natural in quantum mechanical applications. It is natural to denote the result of applying a rotation $R$ or a permutation $P$ to an eigenfunction $\psi$ by $R\psi$ or $P\psi$. It is not necessary to introduce a new letter $A$, $R$ or $P$ for the linear transformation representing the group element $a$, $R$ or $P$.

Another advantage of the notation $av$ is that all notions and theorems of group theory can be directly applied to our spaces $\mathscr{V}$, considered as additive groups with operators. Thus the notion of operator-isomorphism, applied to two vector spaces $\mathscr{V}$ and $\mathscr{V}'$ on which the same group $G$ operates, yields at once the notion *equivalence* of representations:

*Two representations ϱ and ϱ' of a group $\mathscr{G}$ are equivalent, if the vector spaces $\mathscr{V}$ and $\mathscr{V}'$, considered as groups with operators a (elements of $\mathscr{G}$) and β (complex numbers) are operator-isomorphic.*

Or, what amounts to the same thing:

*Two representations of $\mathscr{G}$ are equivalent if the basic vectors in $\mathscr{V}$ and $\mathscr{V}'$ can be chosen in such a way that any group element $a$ is represented, in both representations, by the same matrix $A$.*

If a new basis is introduced in $\mathscr{V}'$, the matrices of the second representation become $P^{-1}AP$. Hence two matrix representations $\varrho(a \to A)$ and $\varrho'(a \to A')$ are equivalent if and only if a fixed matrix $P$ exists such that

$$A' = P^{-1}AP \quad \text{for all } a.$$

The notion of a permissible subgroup $\mathscr{V}$ of an additive group $\mathscr{U}$ at once yields the notion of an *invariant subspace* of $\mathscr{U}$, that is, of a linear subspace $\mathscr{V}$ which is transformed into itself by the group $\mathscr{G}$. If such an invariant subspace exists, which contains more than only the zero vector, but which is not the whole space $\mathscr{U}$, then the representation is called *reducible*, and the vector space $\mathscr{U}$ too is called reducible (with respect to the group $\mathscr{G}$).

What is the form of the matrices of a reducible representation? We may choose a set of basic vectors $u_1, ..., u_h$ in the subspace $\mathscr{V}$, and extend it to a set of basis vectors $u_1, \ldots, u_n$ of the whole space $\mathscr{U}$. Now we have

(11.1)

$$au_\mu = \sum_1^h u_\lambda p_{\lambda\mu} \quad (\mu = 1, \ldots, h)$$

$$au_\nu = \sum_1^h u_\lambda q_{\lambda\nu} + \sum_{h+1}^n u_\lambda s_{\lambda\nu} \quad (\nu = h+1, \ldots, n)$$

Hence the matrix $A$ has the form

$$A = \begin{pmatrix} P & Q \\ 0 & S \end{pmatrix},$$

in which $P$, $Q$, $S$ are arbitrary matrices, whereas 0 is the zero matrix. The matrices $P$ belong to the representation of $\mathscr{G}$ in the subspace $\mathscr{V}$, whereas the matrices $S$ belong to the representation of $\mathscr{G}$ in the factor space $\mathscr{U}/\mathscr{V}$.

The choice of the basic vectors $u_{h+1}, \ldots, u_n$ is to a certain degree arbitrary, the only condition being that $u_1, \ldots, u_n$ are linearly independent and hence form a basis of $\mathscr{U}$. If it is possible, by a suitable choice of those $n-h$ vectors, to make the matrix elements $q_{\lambda\nu}$ all zero, this would mean that the vectors $u_{h+1}, \ldots, u_n$ generate another invariant subspace $\mathscr{W}$. In this case, the whole space $\mathscr{U}$ is the *direct sum* of $\mathscr{V}$ and $\mathscr{W}$:

(11.2)                                  $\mathscr{U} = \mathscr{V} \oplus \mathscr{W}$

The form of the matrices $A$ is now

(11.3)                                  $A = \begin{pmatrix} P & 0 \\ 0 & S \end{pmatrix}$

and the representation is said to be the *sum* of the representations $\varrho'(a \to P)$ and $\varrho''(a \to S)$:

$$\varrho = \varrho' + \varrho''.$$

If this is the case, the matrices $S$ belong to the representation of $\mathscr{G}$ in the subspace $\mathscr{W}$. But we have also seen that they belong to the representation of $\mathscr{G}$ in the factor space $\mathscr{U}/\mathscr{V}$. Hence $\mathscr{W}$ and $\mathscr{U}/\mathscr{V}$ are isomorphic, and we have the *Isomorphy Theorem*:

$$\mathscr{U} = \mathscr{V} \oplus \mathscr{W} \quad implies \quad \mathscr{W} \cong \mathscr{U}/\mathscr{V}.$$

This Isomorphy Theorem is a special case of a more general group-theoretical Isomorphy Theorem, which says that for any normal divisor $\mathcal{V}$ and any subgroup $\mathcal{W}$ of a group $\mathcal{G}$ the isomorphy

$$\mathcal{V}\mathcal{W}/\mathcal{V} \cong \mathcal{W}/\mathcal{D}$$

holds, $\mathcal{V}\mathcal{W}$ being the group consisting of all products $vw$ ($v$ in $\mathcal{V}$ and $w$ in $\mathcal{W}$) and $\mathcal{D}$ being the intersection of $\mathcal{V}$ and $\mathcal{W}$. In the case of additive groups, we have to write $\mathcal{V} + \mathcal{W}$ instead of $\mathcal{V}\mathcal{W}$.

In what follows, we shall mainly consider representations of groups by *unitary* transformations. Any unitary transformation which leaves invariant a subspace $\mathcal{V}$ of $\mathcal{U}$ also leaves invariant the totally orthogonal subspace $\mathcal{W}$, and it follows that the whole space $\mathcal{U}$ is the direct sum of $\mathcal{V}$ and $\mathcal{W}$. Hence in the unitary case we may always suppose that the matrices of a reducible representation have the form (11.3).

Quite generally, if $\mathcal{A}$ and $\mathcal{B}$ are subgroups of an additive group $\mathcal{G}$, we may form the subgroup $\mathcal{A} + \mathcal{B}$, which consists of all sums $a + b$ ($a$ in $\mathcal{A}$, $b$ in $\mathcal{B}$). If $\mathcal{A}$ and $\mathcal{B}$ have only the zero element in common, it follows that every element of $\mathcal{A} + \mathcal{B}$ can be written in *only one* way as a sum $a + b$. In this case the sum $\mathcal{A} + \mathcal{B}$ is called a *direct sum* and is denoted by $\mathcal{A} \oplus \mathcal{B}$.

Just so, we can form sums of several subgroups

$$\mathcal{S} = \mathcal{B}_1 + \mathcal{B}_2 + \cdots + \mathcal{B}_h .$$

If every element of $\mathcal{S}$ can be written in *only one* way as a sum

$$s = b_1 + b_2 + \cdots + b_h ,$$

we have a *direct sum*

$$\mathcal{S} = \mathcal{B}_1 \oplus \mathcal{B}_2 + \cdots \oplus \mathcal{B}_h .$$

A necessary and sufficient condition for this is, that every $\mathcal{B}_j$ has only the zero element in common with the sum of the preceding $\mathcal{B}$'s.

In a multiplicative group, we can form products and direct products of normal divisors $\mathcal{B}$. In this case, the condition for directness is that every $\mathcal{B}$ has only the unit element in common with the product of the preceding $\mathcal{B}$'s.

An additive group with operators is called *irreducible* or *minimal*, if it has no admissible subgroup except itself and the subgroup consisting of zero only. The group is called *completely reducible*, if it is a direct sum of irreducible subgroups:

$$\mathcal{G} = \mathcal{B}_1 \oplus \cdots \oplus \mathcal{B}_h .$$

Just so, a representation is called completely reducible, if its vector space, considered as an additive group with operators, is completely reducible. In this case the representation is a sum of irreducible representations

$$\varrho = \varrho_1 + \varrho_2 + \cdots + \varrho_h$$

and the matrices of the representation can be written in the form

$$A = \begin{pmatrix} P_1 & & & & \\ & P_2 & & & \\ & & \cdot & & \\ & & & \cdot & \\ & & & & \cdot \\ & & & & & P_h \end{pmatrix}$$

with zeros outside the matrices $P_1, \ldots, P_h$.

*Example of a completely reducible representation.* Let $\mathcal{G} = \mathcal{S}_3$ be the group of all permutations of the numbers 1, 2, 3. A representation of $\mathcal{G}$ can be obtained by taking three basic vectors $u_1$, $u_2$, $u_3$, and applying the permutations to these basic vectors, e.g.

$$(123)\, u_1 = u_2 ,$$

$$(123)\, u_2 = u_3 ,$$

$$(123)\, u_3 = u_1 .$$

These permutations induce linear transformations of the 3-dimensional vector space $(u_1, u_2, u_3)$:

$$(123)\, (u_1 c_1 + u_2 c_2 + u c_3) = u_2 c_1 + u_3 c_2 + u_1 c_3 .$$

These transformations leave invariant the positive Hermitean form

$$c_1^* c_1 + c_2^* c_2 + c_3^* c_3 ,$$

hence we have a representation of $\mathcal{S}_3$ by *unitary* linear transformations. The representation is reducible, for the vector

$$v = u_1 + u_2 + u_3$$

is invariant for all permutations. The multiples $vc$ form a one-dimensional invariant linear subspace $\mathscr{V}$. The subspace $\mathscr{W}$ orthogonal to $\mathscr{V}$ is also invariant, and the differences

$$w_1 = u_1 - u_2$$

$$w_2 = u_2 - u_3$$

form a basis of $\mathscr{W}$. This two-dimensional subspace $\mathscr{W}$ is irreducible: it has no one-dimensional invariant subspace. Hence $\mathscr{U}$ is completely reducible:

$$\mathscr{U} = \mathscr{V} + \mathscr{W}.$$

The representation of $\mathscr{S}_3$ in $\mathscr{U}$ is the sum of two irreducible representations of degrees 1 and 2. The representation of degree 1 is the "identity representation": every permutation in $\mathscr{S}_3$ leaves $v$ invariant and is represented by the one-row unit matrix (1). In order to obtain the matrices of the representation of degree 2, we apply the permutations of $\mathscr{S}_3$ to the basic vectors $w_1$ and $w_2$, e.g.

$$(12) \, w_1 = u_2 - u_1 = -w_1 \,,$$

$$(12) \, w_2 = u_1 - u_3 = w_1 + w_2 \,.$$

Hence the matrix representing the permutation (12) is

$$\begin{pmatrix} -1 & 1 \\ 0 & 1 \end{pmatrix},$$

and so on.

**Theorem.** *Unitary representations are always completely reducible.*
*Proof.* If the vector space $\mathscr{U}$ is irreducible, it is a direct sum of just one irreducible representation, hence it is completely reducible. If $\mathscr{U}$ is reducible and if $\mathscr{V}$ is an invariant subspace, $\mathscr{U}$ is the direct sum of $\mathscr{V}$ and the totally orthogonal subspace $\mathscr{W}$. For $\mathscr{V}$ and $\mathscr{W}$ we may repeat the argument. Thus we finally arrive at a complete reduction of the given representation.

The representations occurring in the quantum theory of atoms and molecules are always unitary, because every rotation, reflection or permutation of particles leaves invariant the integral

$$N(\psi) = \int \psi^* \psi \, dV$$

extended over the whole space.

The notion of reducibility is important in Quantum Theory, for the following reason:

Suppose the energy levels of a quantum mechanical system, e.g. of the system of electrons of an atom, are known in a certain approximation, in which some terms of the Hamiltonian $H$ are neglected. Afterwards the neglected terms are added as perturbation terms $\varepsilon W$ to the approximate Hamiltonian $H_0$, thus obtaining the full Hamiltonian

$$H = H_0 + \varepsilon W.$$

Now suppose that $H_0$ and $\varepsilon W$ are invariant with respect to a certain group $\mathscr{G}$. The unperturbed as well as the perturbed eigenfunctions at any energy level are linearly transformed by the transformations of $\mathscr{G}$. These representations of $\mathscr{G}$ may be reducible or irreducible. If we let $\varepsilon$ go to zero, it is reasonable to assume that the transformations of the perturbed eigenfunctions induced by $\mathscr{G}$ tend to the transformations of the unperturbed eigenfunctions. Now it is clear that a reducible set of transformations can never tend to an irreducible set. From matrices of the form

$$\begin{pmatrix} P & Q \\ 0 & S \end{pmatrix}$$

we obtain, in the limit, always matrices of the same form, with a zero matrix in the lower left corner. Also, the degree of the representation (the number of rows and columns) remains the same in the limit.

Of course, it can happen that several energy levels become equal in the limit and that we get a reducible representation for $\varepsilon = 0$, but it never happens that a reducible representation becomes irreducible in the limit.

Hence if we have, in the limit $\varepsilon = 0$, an irreducible representation of degree $n$, the representation will remain irreducible for small $\varepsilon$. We can go on in the same way, making $\varepsilon$ larger and larger. Thus we obtain, also for large perturbations, the result:

*If we have, in a space of eigenfunctions of the unperturbed Hamiltonian $H_0$, an irreducible representation of a group $\mathscr{G}$ which leaves invariant the total Hamiltonian $H = H_0 + \varepsilon W$, the perturbation never gives rise to a splitting of the eigenvalue $E$. After the perturbation we still have an irreducible representation of the same degree $n$.*

Just so, one proves:

*If we have for $\varepsilon = 0$ a completely reducible representation of degree $n$, which is the direct sum of $r$ irreducible representations $\varrho_1, \varrho_2, \dots, \varrho_r$, then the energy term $E$ can split into $r$ terms at most. If it splits into $r$ terms, the eigenfunctions of these terms are transformed according to irreducible representations, which tend to $\varrho_1, \dots, \varrho_r$ for $\varepsilon \to 0$.*

We shall determine all possible irreducible representations of the groups in question. In all cases it will be possible to distinguish the irreducible representations by quantum numbers, which are integers or half-integers. If $\varepsilon$ varies by continuity, such a number cannot jump, hence the representations must remain the same for all $\varepsilon$.

## § 12. Representations of Abelian Groups. Examples

In a unitary representation $a \rightarrow A$ of an Abelian group $\mathcal{G}$ all transformations $A$, $B$, ... representing the group elements $a$, $b$, ... commute:

$$ab = ba, \quad \text{hence } AB = BA.$$

Hence, by the last theorem of § 9, the matrices of the transformations can be diagonalized simultaneously. This means: a set of orthogonal basic vectors $v_1, ..., v_n$ exists such that the one-dimensional subspaces $(v_1), ..., (v_n)$ are transformed into themselves by the group $\mathcal{G}$. It follows that *every unitary representation of an Abelian group is completely reducible, its irreducible components being of degree one.*

**Example 1.** Let $\mathcal{G}$ be *cyclic of order n*. This means that the group consists of the powers $1$, $a$, $a^2, ..., a^{n-1}$ of one generating element $a$, the $n$-th power $a^n$ being $1$.

Irreducible representations are of degree one. Let the generator $a$ be represented by the matrix $(\alpha)$ having only one element $\alpha$. It follows that $a^2$ is represented by $(\alpha^2)$, etc., finally $a^n = 1$ by $(\alpha^n)$. Hence $\alpha^n = 1$, i.e. $\alpha$ is an $n$-th root of unity. In the complex field there are just $n$ different $n$-th roots of unity, viz.

$$\alpha = \exp \frac{2\pi \, im}{n} \qquad (m = 0, 1, ..., n-1)$$

Hence a cyclic group of order $n$ has just $n$ irreducible representations. An arbitrary representation is a sum of representations of degree one.

It may happen that in the decomposition a certain irreducible representation occurs several times. For instance, if $a \rightarrow (\alpha)$ is an irreducible representation,

$$a \rightarrow \begin{pmatrix} \alpha & 0 \\ 0 & \alpha \end{pmatrix}$$

is a reducible representation.

The permutations of two equal particles (electrons or nuclei) form a cyclic group of order 2. The generating element $a$ is the permutation (12) that interchanges the particles. The irreducible representations are given by

$$a \rightarrow (+1) \quad \text{and} \quad a \rightarrow (-1).$$

Vectors $v_+$ belonging to the representation $(+1)$ remain invariant if the permutation $a = (12)$ is applied to them, whereas vectors $v_-$ belonging to $(-1)$ change their sign:

$$av_+ = v_+ , \qquad av_- = -v_- .$$

Vectors $v_+$ are called *symmetric*, vectors $v_-$ *antisymmetric*.

As an application, let us consider the spectrum of the Helium atom. The nucleus is considered as a fixed point in space and the spin is neglected. At every energy level, we have a set of symmetric eigenfunctions or a set of antisymmetric ones (or both, but the interaction between the electrons usually causes the symmetric and antisymmetric terms to become different). Thus, the term spectrum of the Helium atom consists of just two term systems, of which one (the singlet system) has symmetric eigenfunctions only, while the other (the triplet system) has antisymmetric eigenfunctions only.

The same thing holds for the group consisting of the spatial inversion

$$x' = -x, \qquad y' = -y, \qquad z' = -z$$

and the identity. Here too we have two kinds of basis vectors $v_+$ and $v_-$. The vectors $v_+$ are said to have the *inversion character* $+1$, the vectors $v_-$ the inversion character $-1$. Thus, the term spectrum of an arbitrary atom consists of two partial systems differing in the value of the inversion character $w = \pm 1$.

**Example 2.** *The axial rotation group $\mathcal{O}_2$.* Let $a_\beta$ be a rotation over an angle $\beta$ about a fixed axis. These rotations form an Abelian group $\mathcal{G}$: *the axial rotation group $\mathcal{O}_2$.* We shall consider continuous representations

$$a_\beta \to A_\beta .$$

Irreducible representations are of degree 1, because the group is Abelian. Hence we may write any irreducible representation as

$$A_\beta = (\chi(\beta))$$

the matrix element $\chi(\beta)$ being a complex-valued function of $\beta$ satisfying the functional equation

$$\chi(\beta + \gamma) = \chi(\beta) + \chi(\gamma) .$$

The continuous solutions of this equation are

$$\chi(\beta) = e^{c\beta} .$$

For a univalued representation we must require

$$\chi(2\pi) = \chi(0),$$

hence $e^{2\pi c}$ must be equal to 1, and $c$ must be an integer multiple of $i$:

$$c = -mi.$$

Thus, the univalued irreducible representations are given by

(12.1)                    $$\chi(\beta) = e^{-im\beta}$$

with $m = 0, \pm 1, \pm 2, \ldots$. Every unitary continuous representation is composed of these irreducible representations.

The main application is to molecule spectra. A two-atom molecule can be considered, in a first approximation, as a system of electrons moving in the field of two nuclei fixed in space. A rotation about the axis connecting the nuclei transforms the eigenfunctions of any energy level into eigenfunctions of the same level. Thus, a unitary representation of the axial rotation group is obtained, which is continuous, because a small rotation changes the eigenfunctions only a little. Any unitary representation is completely reducible, hence the representation can be decomposed into irreducible representations of the form (12.1). The quantum number $m$ distinguishes eigenfunctions of different type, which behave differently under rotations.

The absolute value $|m|$ is usually denoted by $\Lambda$. The terms with $\Lambda = 0$ ($m = 0$) are called $\Sigma$-terms, those with $\Lambda = 1$ ($m = \pm 1$) $\Pi$-terms, those with $\Lambda = 2$ ($m = \pm 2$) $\Delta$-terms, etc. The Greek letters $\Sigma\Pi\Delta\ldots$ correspond to the latin letters $s\,p\,d\ldots$ of atom spectroscopy (see § 6). We shall see presently why the terms with opposite $m$-values (e.g. $m = +1$ and $m = -1$) are not distinguished in the Greek letter notation.

**Example 3.** *The axial rotation-reflection group $\mathcal{R}_2$.*

The Hamiltonian of a two-atom molecule is invariant (in the approximation just considered) under reflections at planes passing through the axis. The axial rotations and reflections form a non-Abelian group: the *axial rotation-reflection group $\mathcal{R}_2$*.

The group $\mathcal{R}_2$ can be generated by all rotations and one reflection. Let $a$, as before, denote a rotation over an angle about the z-axis, and let $s_y$ denote a fixed reflection, e.g. the one defined by

$$x' = x, \quad y' = -y, \quad z' = z.$$

The following group relations hold:

$$a_\beta a_\gamma = a_{\beta+\gamma}, \qquad a_\beta s_y = s_y a_{-\beta}.$$

In the vector space of any continuous unitary representation of $\mathcal{R}_2$ we can first decompose the representation of the axial rotation group into irreducible representations of degree 1, each characterized by a rotation quantum number $m$. That is, we can introduce basic vectors $v_m$, which are multiplied by $e^{-im\beta}$ upon rotations:

$$a_\beta v_m = e^{-im\beta} v_m.$$

Now suppose first that $m$ is positive:

$$m = \Lambda > 0.$$

The reflection $s_y$ transforms $v_\Lambda$ into a vector $v_{-\Lambda}$, for we have

$$a_\beta(s_y v_\Lambda) = s_y a_{-\beta} v_\Lambda = s_y e^{+i\Lambda\beta} v_\Lambda = e^{+i\Lambda\beta}(s_y v_\Lambda).$$

Putting $s_y v_\Lambda = v_{-\Lambda}$, we now have

$$a_\beta v_\Lambda = e^{-i\Lambda\beta} v_\Lambda,$$
$$a_\beta v_{-\Lambda} = e^{+i\Lambda\beta} v_{-\Lambda},$$
$$s_y v_\Lambda = v_{-\Lambda},$$
$$s_y v_{-\Lambda} = s_y s_y v_\Lambda = v_\Lambda.$$

Hence the two-dimensional subspace $\mathcal{S} = (v_\Lambda, v_{-\Lambda})$ is invariant with respect to the group $\mathcal{R}_2$. In this subspace the representation of $\mathcal{R}_2$ is irreducible, for if there were a one-dimensional invariant subspace $\mathcal{S}'$, we could apply the axial rotation group to it and find an eigenvector $v_M$ in the subspace, having the property

$$a_\beta v_M = e^{-iM\beta} v_M.$$

Here, $M$ can only be $+\Lambda$ or $-\Lambda$, and $v_M$ can only be $v_\Lambda$ or $v_{-\Lambda}$, because these are the only eigenvectors of $a_\beta$ in $\mathcal{S}$. Hence the subspace $\mathcal{S}'$ would contain either $v_\Lambda$ or $v_{-\Lambda}$. However, the reflection $s_y$ transforms $v_\Lambda$ into $v_{-\Lambda}$ and vice versa. Hence $\mathcal{S}'$ must contain $v_\Lambda$ *and* $v_{-\Lambda}$, and so $\mathcal{S}'$ is equal to $\mathcal{S}$.

Thus, $\mathscr{S}$ determines an irreducible representation of $\mathscr{R}_2$, in which the rotations $a_\beta$ and the reflection $s_y$ are represented by the matrices

$$\begin{pmatrix} e^{-i\varLambda\beta} & 0 \\ 0 & e^{+i\varLambda\beta} \end{pmatrix} \quad \text{and} \quad \begin{pmatrix} 0 & 1 \\ 1 & 0 \end{pmatrix}.$$

This representation will be denoted by $\varrho_\varLambda$.

The case $\varLambda = 0$ is different. In this case the vectors $v_\varLambda$ and $s_y v_\varLambda = v_{-\varLambda}$ are both invariant with respect to rotations. We may form their sum and difference

$$v_0^+ = v_\varLambda + v_{-\varLambda},$$
$$v_0^- = v_\varLambda - v_{-\varLambda},$$

and find

$$s_y v_0^+ = v_0^+,$$
$$s_y v_0^- = -v_0^-.$$

The vector $v_0^+$ is either zero or it generates a one-dimensional subspace, in which all vectors are invariant under the group $\mathscr{R}_2$:

$$s_y v_0^+ = v_0^+,$$
$$a_\beta v_0^+ = v_0^+.$$

Just so, $v_0^-$ is either zero or it generates a one-dimensional subspace, whose vectors are invariant under rotations, but not under reflections:

$$s_y v_0^- = -v_0^-,$$
$$a_\beta v_0^- = v_0^-.$$

Hence there are, besides the representations $\varrho_\varLambda(\varLambda > 0)$ of degree 2, two more irreducible representations $\varrho_0^+$ and $\varrho_0^-$, both of degree 1. The matrix representing $a_\beta$ is $(+1)$ in both cases, and the matrix representing $s_y$ is $(+1)$ for $\varrho_0^+$ and $(-1)$ for $\varrho_0^-$.

The final result is:

*The irreducible continuous representations of $\mathscr{R}_2$ are*

$$\varrho_0^+, \varrho_0^-, \varrho_1, \varrho_2, \ldots.$$

The application to molecule spectra is as follows. If the quantum number $\varLambda$ is positive, there are always two linearly independent eigenfunctions $\psi_\lambda$ and $\psi_{-\lambda}$ having the same energy. If $\varLambda$ is zero, i.e. if we have a $\Sigma$-term, there are two kinds of eigenfunctions $\psi_0^+$ and $\psi_0^-$. The corresponding terms are called $\Sigma^+$-terms and $\Sigma^-$-terms.

As this example shows, a non-Abelian group $\mathscr{G}$ may have representations of degree 1. However, these representations are never true, for because the group elements do not commute, but the matrices do. To the products $ab$ and $ba$ corresponds the same matrix $(\alpha\beta)$, hence the "*commutators*"

$$ab(ba)^{-1} = ab\,a^{-1}b^{-1}$$

are represented by the unit matrix. These commutators and their products form a subgroup, the *commutator group* of $\mathscr{G}$. Any representation of degree one is a true representation of some Abelian factor group $\mathscr{G}/\mathscr{N}$, the normal divisor $\mathscr{N}$ containing at least the commutator group.

**Example 4.** The symmetric group $\mathscr{S}_n (n > 2)$ is not Abelian. Among the commutators we find

$$(ij)\,(ijk)\,(ij)^{-1}(ijk)^{-1} = (ijk).$$

These 3-cycles generate the alternating group $\mathscr{A}_n$. Hence, every representation of degree 1 of $\mathscr{S}_n$ is a representation of the factor group $\mathscr{S}_n/\mathscr{A}_n$. This factor group is cyclic of order 2. Hence it has only two representations, the first one being the *identical* or *symmetric* representation, in which all permutations are represented by the unit matrix $(1)$, and the second one the *alternating representation* in which the even permutations are represented by $(+1)$, the odd by $(-1)$. All other representations of $\mathscr{S}_n$ are of higher degree.

**Example 5.** *The symmetric group $\mathscr{S}_3$.*

The irreducible representations of $\mathscr{S}_3$ may be determined by the same method by which the representations of $\mathscr{R}_2$ were determined in Example 3. Starting with an arbitrary representation of $\mathscr{S}_3$ in a vector space $\mathscr{V}$, we may first restrict ourselves to the subgroup $\mathscr{A}_3$, which is generated by the 3-cycle (123). Any representation of this cyclic group is completely reducible, the irreducible representations being of degree one. As we have seen in example 1, there are just three irreducible representations of the cyclic group $\mathscr{A}_3$, of order 3, the generating element (123) being represented by one of the matrices $(1)$ or $(\zeta)$ or $(\zeta^*)$, where $\zeta$ and $\zeta^*$ are third roots of 1:

$$\zeta = \exp\left(\frac{2\pi i}{3}\right) \quad \text{and} \quad \zeta^* = \zeta^{-1} = \exp\left(-\frac{2\pi i}{3}\right).$$

If a vector $v_\zeta$ belongs to the root $\zeta$, the permutation (12) transforms $v_\zeta$ into a vector $v_{\zeta^*}$ belonging to the root $\zeta^* = \zeta^{-1}$, for we have

$$(123)\, v_{\zeta'} = (123)\,(12)\, v_\zeta = (12)\,(123)^{-1} v_\zeta = (12)\,\zeta^{-1} v_\zeta = \zeta^{-1} v_{\zeta^*}.$$

The vectors $v_\zeta$ and $v_{\zeta^*}$ determine a space $(v_\zeta, v_{\zeta^*})$, in which an irreducible representation of degree 2 takes place. On the other hand, the vectors $v_1$, which remain invariant under all permutations of $\mathscr{A}_3$, form a linear subspace, in which a representation of the cyclic group of order 2 generated by (12) takes place. This representation can be decomposed into irreducible representations of degree 1, in which (12) is either represented by $(+1)$ or by $(-1)$. These two representations of degree 1 are known already from the preceding example: they are the "symmetric" and the "antisymmetric" representation of $\mathscr{S}_3$. Besides, there is only one irreducible representation of degree 2 in a space like $(v_\zeta, v_{\zeta^*})$. In this representation, the generating group elements (123) and (12) are represented as follows:

$$(123) \to \begin{pmatrix} \zeta & 0 \\ 0 & \zeta^* \end{pmatrix}, \quad (12) \to \begin{pmatrix} 0 & 1 \\ 1 & 0 \end{pmatrix}.$$

Every irreducible representation is equivalent to one of these three, and every representation is completely reducible, as we shall see in § 14.

The irreducible representation of degree 2 of $\mathscr{S}_3$ which we have found in § 9 must, of course, be equivalent to the representation in the space $(v_\zeta, v_{\zeta^*})$. It is easy to verify this by direct calculation.

## § 13. Uniqueness Theorems

**Theorem 1.** If $\mathscr{G} = \mathscr{G}_1 \oplus \mathscr{G}_2 \oplus \cdots \oplus \mathscr{G}_r$ is a direct sum of irreducible additive groups with operators and if $\mathscr{H}$ is an arbitrary admissible subgroup of $\mathscr{G}$, we have

$$\mathscr{G} = \mathscr{H} \oplus \mathscr{G}_i \oplus \mathscr{G}_j \oplus \cdots \oplus \mathscr{G}_k$$

in which $i, j, \ldots, k$ form a subset of $\{1, 2, \ldots, r\}$.

*Proof.* We form

$$\mathscr{H}_1 = \mathscr{H} + \mathscr{G}_1,$$
$$\mathscr{H}_2 = \mathscr{H}_1 + \mathscr{G}_2,$$
$$\cdots\cdots\cdots\cdots$$
$$\mathscr{H}_r = \mathscr{H}_{r-1} + \mathscr{G}_r = \mathscr{G}.$$

The meet of $\mathcal{H}$ and $\mathcal{G}_1$ is a normal divisor of $\mathcal{G}_1$ and hence (because $\mathcal{G}_1$ is minimal) either equal to $\mathcal{G}_1$ or to $\{0\}$. In the first case, $\mathcal{G}_1$ is contained in $\mathcal{H}$, hence we have $\mathcal{H}_1 = \mathcal{H}$. In the second case, the sum $\mathcal{H} + \mathcal{G}_1$ is direct and we can write $\mathcal{H}_1 = \mathcal{H} \oplus \mathcal{G}_1$:

In the same way we can rewrite $\mathcal{H}_2 = \mathcal{H}_1 + \mathcal{G}_2$ either as $\mathcal{H}_2 = \mathcal{H}_1$ or as $\mathcal{H}_2 = \mathcal{H}_1 \oplus \mathcal{G}_2$. All sums $\mathcal{H}_{i-1} + \mathcal{G}_i$ are either direct sums, or $\mathcal{G}_i$ can be dropped. Thus, we can finally write $\mathcal{G}$ as a direct sum of $\mathcal{H}$ and some of the $\mathcal{G}_i$.

**Theorem 2.** *If*

$$\mathcal{G} = \mathcal{G}_1 \oplus \mathcal{G}_2 \oplus \cdots \oplus \mathcal{G}_r$$

*and also*

$$\mathcal{G} = \mathcal{G}'_1 \oplus \mathcal{G}_2 \oplus \cdots \oplus \mathcal{G}_r$$

*then* $\mathcal{G}_1 \cong \mathcal{G}'_1$.

*Proof.* The Isomorphy Theorem of §11 implies

$$\mathcal{G}_1 \cong \mathcal{G}/(\mathcal{G}_2 \oplus \cdots \oplus \mathcal{G}_r)$$

and also

$$\mathcal{G}'_1 \cong \mathcal{G}/(\mathcal{G}_2 \oplus \cdots \oplus \mathcal{G}_r).$$

**Theorem 3.** *If* $\mathcal{G} = \mathcal{G}_1 \oplus \cdots \oplus \mathcal{G}_r$ *and* $\mathcal{G} = \mathcal{H}_1 \oplus \cdots \oplus \mathcal{H}_s$ *and if all* $\mathcal{G}_i$ *and* $\mathcal{H}_k$ *are irreducible, then* $r = s$, *and the* $\mathcal{G}_i$ *are in some order isomorphic to the* $\mathcal{H}_k$.

*Proof.* Applying Theorem 1 to $\mathcal{H} = \mathcal{H}_2 \oplus \cdots \oplus \mathcal{H}_s$, one obtains

$$\mathcal{G} = (\mathcal{H}_2 \oplus \cdots \oplus \mathcal{H}_s) + \mathcal{G}',$$

$\mathcal{G}'$ being a sum of some of the $\mathcal{G}_k$. Theorem 2 yields

$$\mathcal{G}' \cong \mathcal{H}_1,$$

hence $\mathcal{G}'$ is irreducible, hence the sum $\mathcal{G}'$ consists of one term $\mathcal{G}_i$ only. Re-numbering the $\mathcal{G}_i$, we may assume $\mathcal{G}' = \mathcal{G}_1$.
Hence $\mathcal{G}_1 \cong \mathcal{H}_1$ and

$$\mathcal{G} = \mathcal{H}_2 \oplus \cdots \oplus \mathcal{H}_s \oplus \mathcal{G}_1.$$

Now one can repeat the same argument, applying Theorem 1 to $\mathcal{H} = \mathcal{H}_3 \oplus \cdots \oplus \mathcal{H}_s + \mathcal{G}_1$ and re-numbering $\mathcal{G}_2, \ldots, \mathcal{G}_r$, thus obtaining $\mathcal{G}_2 = \mathcal{H}_2$ and

$$\mathcal{G} = \mathcal{H}_3 \oplus \cdots \oplus \mathcal{H}_s \oplus \mathcal{G}_1 \oplus \mathcal{G}_2,$$

and so on until one obtains

$$\mathcal{G} = \mathcal{H}_s \oplus \mathcal{G}_1 \oplus \cdots \oplus \mathcal{G}_{s-1}.$$

Theorem 2 yields $\mathscr{H}_s \cong \mathscr{G}_s \oplus \cdots \oplus \mathscr{G}_r$. Since $\mathscr{H}_s$ is irreducible, the sum $\mathscr{G}_s \oplus \cdots \oplus \mathscr{G}_r$ can only contain one term, hence we have $r = s$ and $\mathscr{G}_s \cong \mathscr{H}_s$. Thus, Theorem 3 is proved.

**Corrolary.** If a representation $\varrho$ of any group is completely reducible

$$\varrho = \varrho_1 + \cdots + \varrho_r,$$

the irreducible components $\varrho_1, \ldots, \varrho_r$ are uniquely determined but for equivalence. Thus, it is meaningful to say that a completely reducible representation $\varrho$ contains an irreducible component $\varrho_1$ just once, another component $\varrho_2$ three times etc.

Theorem 1 implies

$$\mathscr{G}/\mathscr{H} = \mathscr{G}_p \oplus \mathscr{G}_q \oplus \ldots,$$

the indices $p, q, \ldots$ forming the complementary set to the indices $i, j, \ldots, k$. Now, since every homomorphic image of $\mathscr{G}$ is isomorphic to a certain factor group $\mathscr{G}/\mathscr{H}$, we obtain

**Theorem 4.** *Every homomorphic image of a completely reducible additive group is isomorphic to the direct sum of some components in the decomposition of $\mathscr{G}$.*

## § 14. Kronecker's Product Transformation

From now on, we shall drop the convention that vectors and linear transformations are printed in fat type. As a consequence of this we shall denote the linear transformation $A$ and its matrix $A$ by the same letter $A$. This will not cause any confusion.

Let a linear transformation $A$ of an $m$-dimensional vector space $\mathscr{U}$ and a linear transformation $B$ of an $n$-dimensional vector space $\mathscr{V}$ be given. For instance, we may take as $\mathscr{U}$ the set of all linear forms $u_1 c_1 + \cdots + u_m c_m$ in $m$ indeterminates $u_1, \ldots, u_n$, and just so as $\mathscr{V}$ the set of all forms $v_1 d_1 + \cdots + v_n d_n$ in $n$ indeterminates $v_1, \ldots, v_n$. The indeterminates $u_i$ and $v_k$ are the basic vectors of the two spaces.

The $m \cdot n$ products $u_i v_k$ may be considered as basic vectors of a vector space $\mathscr{W}$. Elements of $\mathscr{W}$ are the bilinear forms

$$\Sigma u_i v_k c_{ik}$$

If the $u_i$ are transformed according to the linear transformation $A$:

$$A u_j = u_j' = \Sigma u_i a_{ij}$$

and the $v_k$ according to the linear transformation $B$:

$$Bv_l = v_l' = \Sigma v_k b_{kl}$$

one obtains for the product $u_i v_k$ a linear transformation too:

(14.1) $$u_j' v_l' = (\Sigma u_i a_{ij})(\Sigma v_k b_{kl}) = \Sigma u_i v_k a_{ij} b_{kl}.$$

This linear transformation in the product space $\mathcal{W}$ is called the *Product Transformation* $A \times B$. The matrix elements $c_{ik, jl}$ of the product transformation are

$$c_{ik, jl} = a_{ij} b_{kl}.$$

It is easy to see that the transformation $A \times B$ can be applied to any product $uv$ by applying $A$ to $u$ and $B$ to $v$:

(14.2) $$(A \times B)(uv) = (Au)(Bv).$$

Now let $a \to A$ and $a \to A'$ be two representations $\varrho$ and $\varrho'$ of a group $\mathcal{G}$. We shall see presently that $a \to A \times A'$ is again a representation: the *product representation* $\varrho\varrho'$.

Just as the vector spaces $\mathcal{U}$ and $\mathcal{V}$ may be regarded as additive groups with operators from $\mathcal{G}$, just so the product space $\mathcal{W}$ generated by the products $uv$ can be regarded as an additive group with operators from $\mathcal{G}$. According to (14.2) we have the following multiplication rule

(14.3) $$a(uv) = (au)(av),$$

in words: *The operator $a$ is applied to a product of vectors by applying it to both factors.*

When $ab$ is applied to $uv$, the result is the same as when first $b$ and next $a$ is applied:

$$(ab)(uv) = a(b(uv)),$$

hence $\varrho\varrho'$ is in fact a representation of $\mathcal{G}$.

In the same way three or more representations can be multiplied, and we have

$$\varrho\varrho'\varrho'' = (\varrho\varrho')\varrho'' = \varrho(\varrho'\varrho'').$$

The proof of the following *theorem* will be left to the reader: *Any product of unitary representations is unitary.*

Now let $\varrho$ and $\varrho'$ be two *irreducible* representations. We ask: Under what conditions does the vector space $\mathcal{W}$ of the product representation $\varrho\varrho'$ contain an invariant vector $w$ different from zero?

Every vector $w$ of $\mathcal{W}$ can be written as

$$w = \Sigma u_i v_k c_{ik} = \sum_1^m u_i v_i'$$

with $m =$ dimension of $\mathcal{U}$, and

$$v_i' = \Sigma v_k c_{ik}.$$

If $w$ is not zero, the $v_i'$ are not all zero. The condition for $w$ to be invariant is

$$aw = w$$

or

(14.4) $$\Sigma(au_i)(av_i') = \Sigma u_j v_j'.$$

Now let

$$au_i = \Sigma u_j \alpha_{ji}.$$

Substituting this into (14.4), one obtains

$$\Sigma u_j \alpha_{ji} \cdot av_i' = \Sigma u_j v_j'$$

or, since the $u_j$ are linearly independent,

(14.5) $$\Sigma \alpha_{ji} \cdot av_i' = v_j'.$$

Now, if $(\alpha_{ij}')$ denotes the transposed matrix $A'$:

$$\alpha_{ij}' = \alpha_{ji}$$

and if $B = (\beta_{hi})$ denotes the inverse matrix of $(\alpha_{ij}')$, we can solve (14.5) for $av_i'$:

(14.6) $$av_i' = \Sigma v_h' \beta_{hi}.$$

This formula implies: $(v_1', ..., v_m')$ is an invariant subspace of $\mathcal{V}$, hence, because $\mathcal{V}$ is irreducible,

$$(v_1', ..., v_m') = \mathcal{V}$$

and hence $m \geq n$. By interchanging the roles of $\mathcal{U}$ and $\mathcal{V}$, one obtains $n \geq m$, hence $m = n$. It follows that $v_1', ..., v_m'$ are linearly independent: they form a basis for $\mathcal{V}$ and may be renamed $v_1, ..., v_n$. Now, by (14.6), the matrix representing $a$ in the representation $\varrho'$ is

$$B = (\beta_{hi}).$$

Hence:

*If the vector space $\mathscr{W}$ of the representation $\varrho\varrho'$ contains an invariant vector w, it is possible to choose such a basis $(v_1, \ldots, v_n)$ in the space of the representation $\varrho'$ that the matrices of $\varrho'$ are just the inverse transposed matrices of $\varrho$, and that*

$$w = \Sigma u_j v_j$$

*holds.*

To every representation $\varrho$ we have just one *dual representation* $\tilde{\varrho}$ such that $w = \Sigma u_j v_j$ is an invariant vector in the product space $\mathscr{W}$. The matrices of $\tilde{\varrho}$ are the inverse transposed matrices to those of $\varrho$. The dual representation of $\tilde{\varrho}$ is again $\varrho$. If $\varrho$ is irreducible, so is $\tilde{\varrho}$, and vice versa.

If $\varrho$ is a unitary representation, and if the basic vectors $u_i$ of $\mathscr{U}$ are orthogonal and normed, the inverse transposed matrices $B = (\beta_{hi})$ are complex conjugate to $A$:

$$B = A^*.$$

Hence the dual of a unitary representation is the complex conjugate representation $a \rightarrow A^*$.

From now on we shall consider only unitary representations. As we have seen, any representation is completely reducible:

(14.7)                              $$\varrho = \varrho_1 + \cdots + \varrho_h.$$

If in the space of the representation $\varrho$ an invariant vector w exists, one of the irreducible components $\varrho_k$ is the identity representation $a \rightarrow 1$, and vice versa.

Now let $\sigma$ be an irreducible unitary representation. From (14.7) and $\sigma$ we may form the product representation

$$\varrho\sigma = \varrho_1\sigma + \cdots + \varrho_h\sigma.$$

If this completely reducible representation contains the identity representation $\iota$ as an irreducible component, one of the products $\varrho_k\sigma$ contains $\iota$, hence this one $\varrho_k$ is equivalent to the dual $\tilde{\sigma}$ of $\sigma$. Conversely, if one of the $\varrho_k$ is equivalent to $\tilde{\sigma}$, the product representation $\varrho\sigma$ contains the identity $\iota$ as an irreducible component.

If we take for $\varrho$ a product representation $\varrho'\varrho''$, we obtain:

*A product representation $\varrho'\varrho''$ contains $\tilde{\sigma}$ as an irreducible component if and only if $\varrho'\varrho''\sigma$ contains $\iota$, the identity representation.*

For representations of degree 1 the product representations are rather trivial: if $\varrho$ is the representation $a \rightarrow \chi(a)$, and if $\varrho'$ is the representation $a \rightarrow \chi'(a)$, the product representation is

$$a \rightarrow \chi(a)\, \chi'(a)\,.$$

If only $\varrho$ is of degree 1: $a \rightarrow \chi(a)$, and if $\varrho'$ is arbitrary: $a \rightarrow A$, the product representation is

$$a \rightarrow \chi(a)\, A\,.$$

Now if $\varrho'$ is reducible, so is $\varrho\varrho'$, and vice versa, for if all matrices of a reducible set are multiplied by numbers $\chi(a)$ or $\chi(a)^{-1}$, the set remains reducible. Hence if $\varrho$ is of degree 1 and $\varrho'$ irreducible, $\varrho\varrho'$ will be irreducible. On the other hand, if $\varrho$ and $\varrho'$ are both of higher degree than 1, the product $\varrho\varrho'$ may well be reducible.

**Examples.** We shall calculate the products of the irreducible representations

$$\varrho_0^+, \varrho_0^-, \varrho_1, \varrho_2, \ldots$$

of the axial rotation-and-reflection group $\mathscr{R}_2$.

First consider the product $\varrho_\lambda \varrho_\mu$, supposing $\lambda$ and $\mu$ to be positive. The basic vectors of the vector space of $\varrho_\lambda$ are $u_{\pm\lambda}$, and those of $\varrho_\mu$ are $v_{\pm\mu}$. The products are

$$u_\lambda v_\mu,\ u_{-\lambda} v_{-\mu},\ u_\lambda v_{-\mu},\ u_{-\lambda} v_\mu\,.$$

The reflection $s_y$ interchanges the first two and also the last two products. If rotations $a_\beta$ are applied to the first pair, the vectors $u_\lambda v_\mu$ and $u_{-\lambda} v_{-\mu}$ are multiplied by

$$e^{-i(\lambda+\mu)\beta} \quad \text{and} \quad e^{i(\lambda+\mu)\beta}$$

respectively; hence the pair is transformed according to $\varrho_{\lambda+\mu}$. Just so, $u_\lambda v_{-\mu}$ and $u_{-\lambda} v_\mu$ are multiplied by

$$e^{-i(\lambda-\mu)\beta} \quad \text{and} \quad e^{i(\lambda-\mu)\beta}\,.$$

Hence, if $\lambda$ and $\mu$ are different, this pair is transformed according to $\varrho_{|\lambda-\mu|}$. On the other hand, if $\lambda = \mu$, the two vectors $u_\lambda v_{-\lambda}$ and $u_{-\lambda} v_\lambda$ are invariant with respect to $a_\beta$. Their sum

$$u_\lambda v_{-\lambda} + u_{-\lambda} v_\lambda$$

is invariant under $s_y$, whereas the difference

$$u_\lambda v_{-\lambda} - u_{-\lambda} v_\lambda$$

is multiplied by $-1$ upon application of $s_y$. Hence we have, provided $\lambda$ and $\mu$ are both positive,

$$\varrho_\lambda \varrho_\mu = \varrho_{\lambda+\mu} + \varrho_{|\lambda-\mu|}, \quad \text{if} \quad \lambda \neq \mu,$$
$$\varrho_\lambda \varrho_\lambda = \varrho_{\lambda+\lambda} + \varrho_0^+ + \varrho_0^-.$$

Next consider the case $\lambda > 0$, and $\mu = 0^+$ or $0^-$. If $\mu$ is $0^+$, the products $u_\lambda v_0$ and $u_{-\lambda} v_0$ are transformed as $u_\lambda$ and $u_{-\lambda}$, hence the product representation is $\varrho_\lambda$:

$$\varrho_\lambda \varrho_0^+ = \varrho_\lambda.$$

If $\mu$ is $0^-$, the products $u_\mu v_0$ and $-u_{-\mu} v_0$ are transformed according to $\varrho_\lambda$, hence

$$\varrho_\lambda \varrho_0^- = \varrho_\lambda.$$

The remaining cases are trivial:

$$\varrho_0^+ \varrho_0^+ = \varrho_0^+,$$
$$\varrho_0^+ \varrho_0^- = \varrho_0^-,$$
$$\varrho_0^- \varrho_0^- = \varrho_0^+.$$

## § 15. The Operators Commuting with all Operators of a Given Representation

Let $\mathscr{U}$ and $\mathscr{V}$ be two vector spaces having a common set of operators $\mathscr{G}$, and let $T$ be a linear transformation mapping $\mathscr{U}$ into $\mathscr{V}$. If $T$ is an operator-homomorphism, this means

(15.1)                    $T(av) = a(Tv)$   for all $a$ in $\mathscr{G}$.

If $\mathscr{G}$ is a group, and if $a \to A$ and $a \to A'$ are representations of $\mathscr{G}$ by linear transformations of $\mathscr{U}$ and $\mathscr{V}$ respectively, we may write (15.1) as

(15.2)                              $TA = A'T$

or, if $\mathscr{U} = \mathscr{V}$ and $A = A'$, as

$$TA = AT.$$

Our problem is, to determine all transformations $T$ satisfying (15.1) or (15.2). In studying this problem it is most convenient to avoid the notations $a \to A$ and $a \to A'$ and to write the condition for $T$ in the simpler form (15.1).

We first consider the case in which $\mathscr{U}$ is irreducible. In this case, the set of all $T$ satisfying (15.1) is given by the *Lemma of Schur*. It consists of two parts 1. and 2.:

1. *If $\mathscr{U}$ is irreducible, any operator-homomorphism $T$ is either an isomorphism or it maps every vector upon zero.*

The very simple proof will be given presently. Let us first discuss, what happens in the two cases. In the first case, $\mathscr{U}$ is operator-isomorphic to an admissible subspace of $\mathscr{V}$. If $\mathscr{V}$ too is irreducible, this admissible subspace must be the whole of $\mathscr{V}$, and $\mathscr{U} \cong \mathscr{V}$. In the special case $\mathscr{U} = \mathscr{V}$, we may use the same set of basic vectors for $\mathscr{U}$ and $\mathscr{V}$. In this case we may assert even more:

2. *The matrix of $T$ is a multiple $\lambda I$ of the identity matrix $I$.*

*Proof.* According to the Homomorphy Theorem, $T$ is an isomorphism of a factor space $\mathscr{U}/\mathscr{W}$. If $\mathscr{U}$ is irreducible, $\mathscr{W}$ is either $\{0\}$ or the whole of $\mathscr{U}$. In the first case, $T$ is an isomorphism. In the second case, $T$ maps every vector upon 0. Thus, 1. is proved.

Now suppose $\mathscr{U} = \mathscr{V}$; we have to prove $T = \lambda I$. Let $\lambda$ be any root of the secular equation

$$\mathrm{Det}\,(T - \lambda I) = 0\,.$$

If $T$ commutes with all operators $a$, so does $T - \lambda I$. Now $T - \lambda I$ has determinant zero, hence it cannot be an isomorphism. Hence, by 1., $T - \lambda I$ must be zero. This proves 2.

Of course, the same assertion $T = \lambda I$ also holds if $\mathscr{U}$ and $\mathscr{V}$ are not supposed to coincide, but only to be isomorphic, provided the basic vectors of $\mathscr{V}$ are just the maps of the basic vectors of $\mathscr{U}$ under the given isomorphism.

By Schur's lemma, the set of all matrices $T$ commuting with all matrices of an *irreducible* representation is completely determined: it consists of all matrices $\lambda I$.

Our next problem is, to determine all matrices $T$ commuting with a *completely reducible* representation.

Let $\mathscr{V}$ be a vector space with a set of operators $\mathscr{G}$, and let $\mathscr{V}$ be completely reducible:

(15.3) $$\mathscr{V} = \mathscr{V}_1 \oplus \mathscr{V}_2 + \cdots \oplus \mathscr{V}_r\,,$$

the $\mathscr{V}_i$ being irreducible. We have to determine all operator-homomorphisms $T$ of $\mathscr{V}$ into $\mathscr{V}$.

In (15.3), we may suppose that $\mathscr{V}_1, \ldots, \mathscr{V}_k$ (say) are isomorphic, but not isomorphic to the other $\mathscr{V}_i$. In $\mathscr{V}_1$ we may choose any set of basic vectors, and in $\mathscr{V}_2, \ldots, \mathscr{V}_k$ we may introduce corresponding sets, which are obtained from the basic vectors in $\mathscr{V}_1$ by isomorphisms mapping $\mathscr{V}_1$ upon $\mathscr{V}_2, \ldots, \mathscr{V}_k$. If this is done, the elements of $\mathscr{G}$ are represented by just the same matrices in $\mathscr{V}_1$, in $\mathscr{V}_2, \ldots$, and in $\mathscr{V}_k$.

The operator-isomorphism $T$ is completely known, as soon as we know how it operates on the vectors of $\mathscr{V}_1$, of $\mathscr{V}_2, \ldots$, of $\mathscr{V}_r$. Now let $v$ be any vector of $\mathscr{V}_1$. Its map $Tv$ can be decomposed according to (15.3):

(15.4)                           $Tv = w = w_1 + w_2 + \cdots + w_r .$

The mapping $v \to w$ is an operator-homomorphism, and so is the mapping $w \to w_1$, hence $v \to w_1$ is an operator-homomorphism. The same holds for the mappings $v \to w_2, \ldots, v \to w_r$. By Schur's lemma, each of these mappings $v \to w_\nu$ is either an isomorphism or a mapping of $\mathscr{V}_1$ upon zero. Isomorphisms are only possible for $\nu = 1, \ldots, k$, for only $\mathscr{V}_1, \ldots, \mathscr{V}_k$ are isomorphic to $\mathscr{V}_1$. All other $w_{k+1}, \ldots, w_r$ must be zero. Moreover, the matrices of the linear transformations $v \to w_1, \ldots, v \to w_k$ must all be multiples of the unit matrix $I$. Thus, the matrix of the mapping $v \to w$ of $\mathscr{V}_1$ into $\mathscr{V}_\mu$ may be written as $\tau_{\mu 1} I$, where $\mu$ goes from 1 to $h$. The same holds for the mappings $\tau_{\mu 2} I$ of $\mathscr{V}_2$ into $\mathscr{V}_\mu$, etc. up to $\mathscr{V}_k$. With $\mathscr{V}_{k+1}$ a new set of isomorphic $\mathscr{V}$'s begins, and so on.

Applying the transformation $T$ to the basic vectors of $\mathscr{V}_1, \mathscr{V}_2, \ldots, \mathscr{V}_r$, we finally obtain the matrix of $T$ in the following form.

$$(15.5) \quad \begin{pmatrix} \begin{array}{ccc} \tau_{11}I & \cdots & \tau_{1k}I \\ \vdots & & \vdots \\ \tau_{k1}I & & \tau_{kk}I \end{array} & & \\ & \begin{array}{c} \tau_{k+1,k+1}I \cdots \\ \vdots \\ \cdots\cdots\cdots \end{array} & \\ & & \ddots \end{pmatrix}$$

The result just obtained can also be stated as follows: If the basic vectors of the isomorphic spaces $\mathscr{V}_1, \ldots, \mathscr{V}_k$ are written below each other in $k$ rows:

$$v_{11}, v_{12}, \ldots, v_{1m} \quad \text{(Basis of } \mathscr{V}_1)$$

$$v_{21}, v_{22}, \ldots, v_{2m} \quad \text{(Basis of } \mathscr{V}_2)$$

$$v_{k1}, v_{k2}, \ldots, v_{km} \quad \text{(Basis of } \mathscr{V}_k),$$

then the operators of $\mathscr{G}$ transform the *rows* of the rectangle into themselves, and all the rows in the same way, whereas the operator $T$ transforms the *columns* of the rectangle into themselves, all by the same linear transformation with matrix

(15.6)
$$T_1 = \begin{pmatrix} \tau_{11} \, \tau_{12} \cdots \tau_{ik} \\ \vdots \\ \tau_{k1} \, \tau_{k2} \cdots \tau_{kk} \end{pmatrix}$$

The other subspaces $V_{k+1}, \ldots, V_r$ give rise to similar rectangles of basic vectors.

It is easy to see that the transformations $T$ defined by matrices like (15.5) do commute with the operators $a$ of $\mathscr{G}$. Hence the problem of determining all these commuting transformations $T$ is completely solved. The coefficients $\tau_{\mu\nu}$ are arbitrary.

At this stage it is useful to introduce some notions from the theory of algebras.

**Definition 1.** An *algebra* is a ring $\mathscr{R}$ which is at the same time a vector space with respect to a field $\mathscr{F}$. This means: for the elements $u, v, \ldots$ of $\mathscr{R}$ three operations are defined: addition $u + v$, multiplication $uv$, and multiplication $v\gamma$ of vectors $v$ by elements $\gamma$ of the field $\mathscr{F}$. In this book, $\mathscr{F}$ is always the field $\mathbb{C}$ of complex numbers, and $\mathscr{R}$ is always finite-dimensional.

**Example 1.** The *group algebra* $\mathscr{R}$ of a finite group $\mathscr{G}$ is obtained by taking the elements $s, t, \ldots$ of the group as basic elements of the vector space $\mathscr{R}$. The elements of the ring $\mathscr{R}$ are all sums

$$v = \sum_t t\gamma_t$$

with complex coefficients $\gamma_t$. Products $uv$ are defined by

$$(\Sigma s\beta_s)\,(\Sigma t\gamma_t) = \sum_{s,t} st\beta_s\gamma_t \, .$$

**Example 2.** All $n$ by $n$ matrices $T$ with arbitrary complex matrix elements $\tau_{ik}$ form an algebra $\mathscr{M}_n$ of dimension $n^2$: the *full matrix algebra* of degree $n$.

In order to show that $\mathscr{M}_n$ is of dimension $n^2$, we introduce $n^2$ basic elements $C_{ik}$. The matrix $C_{ik}$ has one matrix element 1 in the $i$-th row and $k$-th column, all other matrix elements being zero. Every matrix $T$ is a sum

$$T = \Sigma C_{ik}\tau_{ik} \, .$$

The multiplication rules for the $C_{ik}$ are

$$C_{hi}C_{ik} = C_{hk}$$

$$C_{hi}C_{jk} = 0 \qquad (i \neq j)$$

**Definition 2.** An algebra $\mathscr{R}$ is called a *direct sum of sub-algebras* $\mathscr{A}_1, ..., \mathscr{A}_r$:

$$\mathscr{R} = \mathscr{A}_1 \oplus \cdots \oplus \mathscr{A}_r$$

if and only if:

1. the vector space $\mathscr{R}$ is the direct sum of the subspaces $\mathscr{A}_1, ..., \mathscr{A}_r$,

2. the $\mathscr{A}_i$ are rings,

3. all products $\mathscr{A}_i \mathscr{A}_k$ with $i \neq k$ are zero. This means: $a_i b_k = 0$ for $a_i$ in $\mathscr{A}_i$, $b_k$ in $\mathscr{B}_k$.

If these three conditions are satisfied, the structure of $\mathscr{R}$ is completely determined by the structures of the subrings $\mathscr{A}_i$. In fact, if

$$a = a_1 + \cdots + a_r$$

and

$$b = b_1 + \cdots + b_r$$

are elements of $\mathscr{A}$, their sum and product are

$$a + b = (a_1 + b_1) + \cdots + (a_r + b_r)$$

$$ab = a_1 b_1 + \cdots + a_r b_r.$$

We have determined all transformations $T$ commuting with a completely reducible representation. The matrices $T$ have the form (15.5), the $\tau_{ij}$ being arbitrary complex numbers. To every $T$ corresponds a finite set of matrices $T_1, ..., T_q$, as in (15.6):

$$(15.7) \qquad T_1 = \begin{pmatrix} \tau_{11} & \cdots & \tau_{1k} \\ \vdots & & \\ \tau_{k1} & \cdots & \tau_{kk} \end{pmatrix}, \qquad T_2 = \begin{pmatrix} \tau_{k+1,k+1} & \cdots \\ \vdots & \\ & \end{pmatrix}, ...$$

The matrices $T$ form an algebra $\mathscr{R}$, which is a direct sum of subalgebras $\mathscr{A}_1, ..., \mathscr{A}_q$, defined as follows. An element of $\mathscr{A}_1$ is obtained by assuming $T_2 = 0, ..., T_q = 0$, whereas the $\tau_{ij}$ in $T_1$ are arbitrary. Just so, if $T_2$ is arbitrary and all other $T_i$ zero, on obtains an element of $\mathscr{A}_2$, and so on. It is clear that the algebras $\mathscr{A}_1, ..., \mathscr{A}_s$

are isomorphic to complete matrix algebras of degrees $k_1 = k, k_2, ..., k_q$ and that $\mathscr{R}$ is their direct sum:

$$\mathscr{R} = \mathscr{A}_1 \oplus \cdots \oplus \mathscr{A}_s.$$

Hence we have the *Theorem*:

*The algebra of all matrices T commuting with a completely reducible set of matrices is a direct sum of algebras isomorphic to full matrix algebras of degrees $k_1, ..., k_q$.*

The same proof also yields the following result:

*If two sets of linear transformations A and T are given, such that every A commutes with every T, and if at least one of the sets is completely reducible, it is possible to choose basic vectors ordered in rectangles*

$$
\begin{matrix}
v_{11} & \cdots & v_{1n} \\
\vdots & & \vdots \\
v_{s1} & \cdots & v_{sn}
\end{matrix}
$$

*such that every transformation A transforms the rows into themselves, all rows according to one and the same matrix, whereas every transformation T transforms the columns of the rectangle into themselves, all columns according to one and the same matrix.*

This result will prove very useful in Quantum Mechanics.

## § 16. Representations of Finite Groups [1]

Let $\mathscr{G}$ be a finite group having $h$ elements. In the space of any representation of $\mathscr{G}$ one can choose an arbitrary positive Hermitean form, and apply all transformations of the group to this form. The sum of the transformed forms is again a positive Hermitean form $H$. The transformations representing the group elements leave $H$ invariant, hence they are unitary. This implies:

*Every representation of a finite group $\mathscr{G}$ is completely reducible.*

In the preceeding section we have constructed the group algebra $\mathscr{R}$ of $\mathscr{G}$, the ring of all sums

$$v = \Sigma t\gamma_t$$

with arbitrary complex coefficients $\gamma_t$. This group algebra may be considered as a vector space on which the group $\mathscr{G}$ operates, for if $s$ is

---

[1] The contents of this section is not strictly necessary for the quantum mechanical applications to be discussed in this book. However, it is a must for anyone who wants to understand the central ideas of the theory of Group Representations. The theory was first developed by G. Frobenius. The methods of proof used in this section are due to Emmy Noether.

an arbitrary group element, the product $sv$ is again in $\mathscr{R}$, and the transformation

$$v \to sv$$

is a linear transformation of $\mathscr{R}$ into itself. This particular representation of $\mathscr{G}$ is called the *regular representation* of $\mathscr{G}$. Its degree is equal to the order $h$ of the group.

The invariant subspaces of the vector space $\mathscr{R}$ are those linear subspaces $\mathscr{L}$ which contain, with every element $v$, all products $sv$ as well. If this is the case, $\mathscr{L}$ also contains all products

$$\Sigma \, s\beta_s \cdot v = u \cdot v$$

$u$ being an arbitrary element of the ring $\mathscr{R}$. Such subspaces are called *left ideals* in the ring $\mathscr{R}$.

Since every representation is completely reducible, so is the regular representation. This means: $\mathscr{R}$ *is a direct sum of minimal left ideals*:

$$(16.1) \qquad\qquad \mathscr{R} = \mathscr{L}_1 \oplus \cdots \oplus \mathscr{L}_r \,.$$

**Theorem 1.** *Every irreducible representation of $\mathscr{G}$ is contained in the regular representation, i.e. it is equivalent to the representation defined by a minimal left ideal $\mathscr{L}$.*

*Proof.* We first note that every representation of the group $\mathscr{G}$ defines a representation of the group algebra $\mathscr{R}$, as follows: if the group elements $s$ are represented by matrices $S$, the element $\Sigma \, s\,\beta_s$ is represented by $\Sigma S\beta_s$.

Now let $v$ be a fixed non-zero vector in the vector space $\mathscr{V}$ of an irreducible representation. We consider $\mathscr{R}$ and $\mathscr{V}$ as additive groups with operators from $\mathscr{G}$, and we define a mapping of $\mathscr{R}$ into $\mathscr{V}$ by

$$(16.2) \qquad\qquad\qquad u \to uv \,.$$

This linear mapping is an operator homomorphism, for we have, for every $s$ in $\mathscr{G}$

$$su \to (su)\, v = s(uv) \,.$$

The map of $\mathscr{R}$ under the homomorphism (16.2) contains the vector $e \cdot v = v$, so it is not zero, hence it must be the whole of $\mathscr{V}$, i.e. (16.2) is a homomorphism of $\mathscr{R}$ on to $\mathscr{V}$.

Now we can apply theorem 4 of § 13, which says: Every homomorphic image of a completely reducible additive group is isomorphic to a direct sum of some components in the decomposition of the group. Applying this to the additive group $\mathcal{R}$ and its homomorphic image $\mathcal{V}$ we find: $\mathcal{V}$ is isomorphic to a direct sum of some of the left ideals occurring in the decomposition (16.1). But since $\mathcal{V}$ is irreducible, the direct sum can only consist of just one $\mathcal{L}_i$. Hence $\mathcal{V}$ is isomorphic to one of the minimal left ideals $\mathcal{L}_i$, which proves the theorem.

**Corollary.** Since all representations of $\mathcal{G}$ are completely reducible and since all irreducible representations are equivalent to representations defined by minimal left ideals, it follows that *every representation can be decomposed into irreducible representations equivalent to representations defined by minimal left ideals.* In other words: Every representation space is a direct sum of minimal invariant subspaces, isomorphic to minimal left ideals of $\mathcal{R}$.

In order to find out how many inequivalent irreducible representations there are, we have to determine the structure of the ring $\mathcal{R}$. To this purpose we determine all linear transformations of $\mathcal{R}$ into itself which commute with all matrices of the regular representation. In other words, we determine all operator-homomorphisms of $\mathcal{R}$, considered as an additive group with operators from $\mathcal{G}$. Let $T$ be such an operator-homomorphism. We have for all $s$ in $\mathcal{G}$ and $v$ in $\mathcal{R}$

$$Tsv = sTv$$

and hence

$$T \Sigma s\beta_s v = \Sigma s\beta_s Tv$$

$$Tuv = uTv$$

for all $u$ and $v$ in $\mathcal{R}$. Now let $v$ be the unit element $e$ of $\mathcal{G}$. We obtain

$$Tu = uTe.$$

Putting $Te = t$, we finally obtain

$$Tu = ut.$$

This means: the linear transformation $T$ is just the multiplication of $u$ by the fixed factor $t$. To every $T$ we have just one $t$, and conversely. To the product $T_1 T_2$ corresponds the product $t_2 t_1$, with interchanged factors, for we have

$$T_1 T_2 \cdot e = T_1(T_2 e) = T_1 t_2 = t_2 t_1.$$

To the sum $T_1 + T_2$ corresponds the sum $t_1 + t_2$:

$$(T_1 + T_2)\, e = T_1 e + T_2 e = t_1 + t_2\,.$$

Hence we have an *anti-isomorphism* between the algebra $\mathscr{A}$ of all operators $T$ and the algebra $\mathscr{R}$. Now we know that $\mathscr{A}$ is a direct sum of full matrix algebras, hence $\mathscr{R}$ is anti-isomorphic to a direct sum of full matrix algebras. Now a full matrix algebra is anti-isomorphic to itself, for if $A'$ is the transpose of the matrix $A$, the mapping $A \to A'$ is an anti-isomorphism:

$$(AB)' = B'A'\,.$$

Hence we have

**Theorem 2.** *The group algebra $\mathscr{R}$ of a finite group is isomorphic to a direct sum of full matrix algebras:*

$$(16.3) \qquad\qquad \mathscr{R} = \mathscr{R}_1 \oplus \cdots \oplus \mathscr{R}_q\,.$$

In order to obtain all irreducible representations of $\mathscr{R}$, we have to find a decomposition (16.1) of $\mathscr{R}$ into left ideals. For this purpose, it is sufficient to decompose every single matrix algebra $\mathscr{R}_i$ into minimal left ideals. Consider e.g. the algebra $\mathscr{R}_1$.

Let $\mathscr{R}_1$ be the full matrix ring $\mathscr{M}_n$. We introduce the $n^2$ basic elements $C_{ik}$ defined in § 15. The elements

$$(16.4) \qquad\qquad C_{11}, C_{21}, \ldots, C_{n1}$$

generate a left ideal $\mathscr{L}_1$ in $\mathscr{M}_n$. Just so, the $C_{i2}$ generate a left ideal $\mathscr{L}_2$, etc. Thus we obtain $n$ left ideals $\mathscr{L}_1, \ldots, \mathscr{L}_n$. In order to find the representation of $\mathscr{M}_n$ defined by the left ideal $\mathscr{L}_1$, one has to multiply an arbitrary element

$$(16.5) \qquad\qquad t = \Sigma\, C_{ik}\, \tau_{ik}$$

by the basic elements (16.4) and express the products by the same basic elements:

$$\left(\sum_{i,k} C_{ik}\, \tau_{ik}\right) C_{j1} = \Sigma\, C_{ij}\, \tau_{ij}\, C_{j1}$$

$$= \Sigma\, C_{i1}\, \tau_{ij}\,.$$

Hence the matrix representing the element (16.5) of $\mathscr{M}_n$ is just the matrix $(\tau_{ij})$.

The representation just constructed was defined by the left ideal $\mathscr{L}_1$. The left ideals $\mathscr{L}_2, \ldots, \mathscr{L}_n$ define just the same representation; hence the left ideas $\mathscr{L}_1, \ldots, \mathscr{L}_n$ are operator-isomorphic. The representation $t \to (\tau_{ij})$ is irreducible, because the $\tau_{ij}$ are completely arbitrary complex numbers. Hence the $\mathscr{L}_i$ are minimal ideals and we have the decomposition

$$(16.6) \qquad \mathscr{R}_1 = \mathscr{M}_n = \mathscr{L}_1 \oplus \cdots \oplus \mathscr{L}_n.$$

The same holds for $\mathscr{R}_2, \ldots, \mathscr{R}_q$, only the numbers $n_2, \ldots, n_q$ may be different from $n = n_1$. Inserting the decompositions (16.6) in (16.3), one finally obtains

$$\mathscr{R} = (\mathscr{L}_1 \oplus \cdots \oplus \mathscr{L}_{n_1}) \oplus (\mathscr{L}_{n_1+1} \oplus \cdots \oplus \mathscr{L}_{n_1+n_2}) \oplus \cdots$$

in which $\mathscr{L}_1, \ldots, \mathscr{L}_{n_1}$ are isomorphic, as are $\mathscr{L}_{n_1+1}, \ldots, \mathscr{L}_{n_1+n_2}$, etc. The elements $t$ of $\mathscr{R}_1$ are represented by the matrices $(\tau_{ij})$ in the representation $\varrho_1$ defined by $\mathscr{L}_1$ and by 0 in all other representations $\varrho_2, \ldots, \varrho_q$. The representations $\varrho_1, \ldots, \varrho_q$ are not equivalent. *The number of inequivalent representations $\varrho_i$ is equal to the number of rings $\mathscr{R}_i$ in the decomposition* (16.3). There are no other irreducible representations besides $\varrho_1, \ldots, \varrho_q$. *In the regular representation every representation $\varrho_i$ occurs as often as its degree $n_i$ indicates.* The dimension of $\mathscr{R}_i$ is $n_i^2$, hence the dimension of the whole ring $\mathscr{R}$ is

$$(16.7) \qquad h = \Sigma \, n_i^2.$$

An immediate consequence is the *Theorem of Burnside*:
*Every irreducible representation of degree $n$ has just $n^2$ linearly independent matrices.*

The theorems just proved, including Burnside's theorem, hold not only for the group algebra of a finite group, but more generally for any algebra $\mathscr{R}$ with a unit element $e$, which is *completely reducible*, i.e. which is a direct sum of minimal left ideals. The proofs are just the same. In particular, the following *theorem* holds:

*A single full matrix ring $\mathscr{R}$ has only one irreducible representation, which is given by the matrices themselves.*

**Application: Dirac's Matrices.** Let us consider the problem to find 4 matrices $\Gamma_1$, $\Gamma_2$, $\Gamma_3$, $\Gamma_4$ satisfying the following equations:

$$(16.8) \qquad \Gamma_\alpha^2 = 1, \Gamma_\alpha \Gamma_\beta = -\Gamma_\beta \Gamma_\alpha \quad (i \neq k).$$

By means of these equations, all products of matrices can be expressed by the following 16 products:

$$(16.9) \qquad 1, \Gamma_\alpha, \Gamma_\alpha \Gamma_\beta, \Gamma_\alpha \Gamma_\beta \Gamma_\delta, \Gamma_1 \Gamma_2 \Gamma_3 \Gamma_4 \quad (\alpha < \beta < \delta)$$

Now if the matrices are not yet known, we may form an algebra having 16 basic elements

(16.10)                          $1, \gamma_\alpha, \gamma_{\alpha\beta}, \gamma_{\alpha\beta\delta}, \gamma_{1234}$     $(\alpha < \beta < \delta)$

which are multiplied just like the 16 products (16.9) according to the multiplication rules (16.8). This algebra is uniquely defined. Every set of 4 matrices $\Gamma_i$ satisfying (16.8) yields a representation of the algebra, and vice versa. Hence our problem is: to find all representations of a definite algebra by matrices.

Now Dirac[2] has shown that a representation by 4-row matrices exists, in which the 16 basic elements (16.10) are represented by 16 linearly independent matrices. Hence our algebra is isomorphic to the full matrix algebra $\mathcal{M}_4$. Hence by the theorem just formulated, every irreducible representation is equivalent to Dirac's. Every reducible representation is found by repeating Dirac's representation a certain number of times.

After this digression, we return to representations of finite groups $\mathcal{G}$. We want to find out, how many inequivalent representations $\mathcal{G}$ has. To solve this problem, we determine the "center" of the group algebra $\mathcal{R}$, i.e. the set of all those elements $\Sigma\, s\beta_s$ which commute with all elements of the algebra. For this, it suffices that they commute with all group elements $t$, which means

or

$$t \cdot \Sigma\, s\beta_s = \Sigma\, s\beta_s \cdot t$$

$$\Sigma\, tst^{-1}\beta_s = \Sigma\, s\beta_s$$

for all $t$. This in turn means that all conjugates $tst^{-1}$ of any group element $s$ have the same coefficient $\beta_s$ in the sum $\Sigma\, s\beta_s$. Now let $u$ be the sum of all different conjugates of $s$. Then any element of the center must be of the form

$$\sum_u u\beta_u .$$

Hence the center of $\mathcal{R}$ is a vector space whose dimension is equal to the number of different classes of conjugate group elements.

On the other hand, the center can also be determined from formula (16.3). Let

$$t = t_1 + \cdots + t_q$$

be the decomposition of any element $t$ of $\mathcal{R}$ according to (16.3). If $t$ belongs to the center of $\mathcal{R}_1$, every $t_i$ must belong to the center of the full matrix ring $\mathcal{R}_i$, and conversely. According to Schur's lemma (§ 15) the center of any full matrix ring $R_i$ consists of multiples $e_i\lambda$ of the unit matrix $e_i$ only. Hence the center of $R$ consists of all sums

$$e_1\lambda_1 + \cdots + e_q\lambda_q .$$

---

[2] P. A. M. Dirac: The quantum theory of the electron. Proc. Roy. Soc. London A**117**, p. 610.

It follows that the dimension of the center is just equal to the number $q$ of inequivalent irreducible representations $\varrho_1, \ldots, \varrho_q$. Hence we have:

*The number of irreducible representations is equal to the number of classes of conjugate group elements in $\mathscr{G}$.*

**Examples.** *1. The Symmetric Group $\mathscr{S}_3$.* The number of elements is 6. The classes of conjugate elements are:

1) the class of $e$, consisting of $e$ only,

2) the class of (123), consisting of (123) and (132),

3) the class of (12), consisting of (12), (13) and (23). Hence there are just 3 representations: the same we have found in § 12, Example 5. Their degrees are 1, 1, 2, and we have indeed

$$6 = 1^2 + 1^2 + 2^2 .$$

*2. The Symmetric Group $\mathscr{S}_4$.* The number of elements is 24. The 5 classes are represented by the elements

$$e, (123), (12)(34), (12), (1234) .$$

A normal divisor $\mathscr{V}_4$ consists of the elements

$$e, (12)(34), (13)(24), (14)(23) .$$

The factor group $\mathscr{S}_4/\mathscr{V}_4$ is isomorphic to $\mathscr{S}_3$, hence it has two representations of degree 1 and one of degree 2. These three representations yield representations of $\mathscr{S}_4$ which are, of course, not true, since all four elements of $\mathscr{V}_4$ are represented by unity.

The equation

$$24 = 1^2 + 1^2 + 2^2 + n_4^2 + n_5^2$$

yields $n_4^2 + n_5^2 = 18$, hence $n_4 = n_5 = 3$. One representation of degree 3 is obtained by taking four vectors $e_1, e_2, e_3, e_4$ in a 3-dimensional vector space of which the first three are linearly independent, the sum of all four being zero:

$$e_1 + e_2 + e_3 + e_4 = 0 .$$

The permutations of the four vector yield linear transformations of the vector space. For instance, if $e_1, e_2, e_3$ are taken as basic vectors, the permutation (12) is represented by the matrix

$$\begin{pmatrix} 0 & 1 & 0 \\ 1 & 0 & 0 \\ 0 & 0 & 1 \end{pmatrix}$$

and so on. Thus we obtain an irreducible representation $\varrho_4$ of degree 3.

The other representation $\varrho_5$ of degree 3 is obtained from $\varrho_4$ by multiplying all matrices representing odd permutations by $-1$.

*3. The Alternating Group $\mathscr{A}_4$.* The order is 12. The 4 classes are represented by the elements

$$e, (123), (132), (12)(34).$$

Hence there are 4 representations. The factor group $\mathscr{A}_4/\mathscr{V}_4$ is cyclic of order 3, hence it has just 3 representions of order 1. The formula

$$12 = 1^2 + 1^2 + 1^2 + n_4^2$$

implies $n_4 = 3$. The missing representation of degree 3 is just the same as any one of the two representations $\varrho_4$ and $\varrho_5$ of $\mathscr{S}_3$ we have just found, applied to even permutations only.


## § 17. Group Characters

The trace of the matrix $(a_{ik})$ of a linear transformation $A$

$$\mathrm{Tr}(A) = \sum_i a_{ii}$$

is an invariant of the transformation: it does not depend on the choice of the basic vectors of the vector space $\mathscr{V}$, as we have seen in § 9. Now let $a \to A$ be a representation $\varrho$ of a group $\mathscr{G}$. We may consider the trace of $A$ as a function of $a$ and call it $\mathrm{Tr}(a)$ or $\mathrm{Tr}_\varrho(a)$:

$$\mathrm{Tr}(a) = \mathrm{Tr}_\varrho(a) = \mathrm{Tr}(A).$$

This trace will be called *the trace of $a$ in the representation $\varrho$*. In the case of an irreducible representation the trace is called a *character* and denoted by $\chi(a)$:

$$\chi(a) = \mathrm{Tr}_\varrho(a) \quad \text{for irreducible } \varrho.$$

Conjugate elements $a$ and $b^{-1} a b$ have the same traces:

$$\mathrm{Tr}(b^{-1}ab) = \mathrm{Tr}(B^{-1}AB) = \mathrm{Tr}(A) = \mathrm{Tr}(a).$$

Hence the traces and characters are *class functions*: they have a fixed value for every class of conjugate group elements.

Calculation of traces and characters is often useful if one wishes to decompose a reducible representation. For this purpose one needs the *Orthogonality Property* of the characters.

Let $s \to A(s)$ and $s \to B(s)$ be two irreducible representations of a finite group $\mathscr{G}$, and let $C$ be the matrix of an arbitrary linear transformation mapping the space $\mathscr{V}$ of the second representation into the space $\mathscr{U}$ of the first one. We now form the matrix

$$P = \sum_t A(t) \, C \, B(t^{-1}),$$

the summation extending over all group elements $t$. The linear transformation $P$ commutes with all group elements $s$:

$$A(s) \, P = A(s) \sum_t A(t) \, C \, B(t^{-1})$$

$$= \sum_t A(st) \, C \, B(t^{-1} s^{-1}) \, B(s) = P B(s).$$

This implies, by Schur's Lemma (§ 15),

$P = 0$, if the representations $A(t)$ and $B(t)$ are not equivalent,

$P = \lambda I$, if the representations are equal.

Inserting the definition of $P$, one obtains

$$\sum_i \sum_j \sum_t a_{hi}(t) \, c_{ij} \, b_{jk}(t^{-1}) = \begin{cases} 0, & \text{if } A(s) \text{ and } B(s) \text{ are not equivalent} \\ \beta \, \delta_{hk}, & \text{if } A(s) = B(s) \end{cases}$$

or, since the $c_{ij}$ are quite arbitrary,

$$(17.1) \quad \sum_t a_{hi}(t) \, b_{jk}(t^{-1}) = \begin{cases} 0, & \text{if } A(s) \text{ and } B(s) \text{ are not equivalent} \\ \beta_{ij} \, \delta_{hk} & \text{if } A(s) = B(s). \end{cases}$$

In order to determine the $\beta_{ij}$ in the case $A(s) = B(s)$, we put $h = k$ and perform a summation over $k$:

$$\sum_t \sum_k a_{jk}(t^{-1}) \, a_{ki}(t) = \beta_{ij} \sum_k \delta_{kk}$$

Since $A(t^{-1}) \, A(t)$ is the unit matrix $(\delta_{ji})$, this equation simplifies to

$$\sum_t \delta_{ji} = \beta_{ij} \sum_k \delta_{kk}$$

or, if $n$ is the degree of the representation and $h$ the order of the group, to

$$h\delta_{ij} = n\beta_{ij}.$$

Solving for $\beta_{ij}$ and substituting into (17.1), we obtain

$$\sum_t a_{hi}(t)\, b_{jk}(t^{-1}) = \begin{cases} 0, \text{ if } A \text{ and } B \text{ are not equivalent} \\ \frac{h}{n}\delta_{ij}\delta_{hk} \text{ if } A = B. \end{cases}$$

If the representation $B(s)$ is unitary, we have

$$B(t^{-1}) = B(t)^{-1} = B(t)^\dagger$$

or

$$b_{jk}(t^{-1}) = b_{kj}^*(t)$$

hence

(17.2)     $$\sum_t a_{hi}(t)\, b_{kj}^*(t) = \begin{cases} 0, \text{ if } A \text{ and } B \text{ are not equivalent} \\ \frac{h}{n}\delta_{ij}\delta_{hk} \text{ if } A = B. \end{cases}$$

The equations (17.2) express the Orthogonality Property of the matrix elements. Putting $h = i$ and $k = j$ and summing over $h$ and $k$, one obtains the Orthogonality Property for the characters:

(17.3)     $$\sum_t \chi^{(1)}(t)\, \chi^{(2)}(t)^* = 0 \quad \text{or} \quad h,$$

namely 0 for the characters of inequivalent representations, $h$ for equivalent representations. If $\chi^{(1)}, \ldots, \chi^{(r)}$ are the characters of a complete set of inequivalent irreducible representations $\varrho_1, \ldots, \varrho_r$, and if any representation $\varrho$ contains the representation $\varrho_i$ just $c_i$ times, the trace of $t$ in the representation $\varrho$ is

$$\mathrm{Tr}_\varrho(t) = \sum_i c_i \chi^{(i)}(t).$$

Multiplying this by $\chi^{(k)}(t)^*$ and applying (17.3), one obtains

(17.4)     $$\sum_t \chi^{(k)}(t)^* \,\mathrm{Tr}(t) = c_k h.$$

By means of this equation, the numbers $c_k$ can be calculated from the traces $\mathrm{Tr}(t)$ and the characters $\chi^{(k)}(t)$. It follows that *a representation*

$\varrho$ *is uniquely determined but for equivalence as soon as the traces* $\mathrm{Tr}(t)$ *of all group elements are known.*

*Application.* Let $\varrho$ and $\sigma$ be any two representations. We want to decompose the product representation $\varrho\sigma$. The trace of a product matrix $A \times B$ with elements

$$c_{ik,jl} = a_{ij}b_{kl}$$

is

$$(17.5) \qquad \sum_{i,k} c_{ik,ik} = \sum_{i,k} a_{ii}b_{kk} = \left(\sum_{i} a_{ii}\right)\left(\sum_{k} b_{kk}\right) = \mathrm{Tr}(A) \cdot \mathrm{Tr}(B).$$

Hence the trace function $\mathrm{Tr}(t)$ of a product representation $\varrho\sigma$ is the product of the trace functions of the representations $\varrho$ and $\sigma$.

**Example.** Let $\iota$ (identity), $\alpha$ (antisymmetrical) and $\varrho$ (representation of degree 2) be the three irreducible representations of $\mathscr{S}_3$. From (17.5) and (17.4) we find

$$\iota \times \iota = \iota \qquad \alpha \times \alpha = \iota \qquad \varrho \times \varrho = \iota + \alpha + \varrho$$

$$\iota \times \alpha = \alpha \qquad \alpha \times \varrho = \varrho$$

$$\iota \times \varrho = \varrho.$$

Chapter III

# Translations, Rotations and Lorentz Transformations

First, the general notion "Lie Group" will be explained. It will be shown that a Lie Group is generated by "infinitesimal transformations", and all representations of the group can be obtained from representations of the infinitesimal "Lie Algebra". Next, the general theory will be applied to translations in time, to three-dimensional rotations and to the Lorentz group.

## § 18. Lie Groups and their Infinitesimal Transformations

### A. Lie Groups

A *Lie Group* is a group and at the same time an $n$-dimensional manifold, which means that in the neighbourhood of any group element $T_0$ the group elements $T(s)$ are determined by $n$ real parameters $s_1, ..., s_n$.

**Example.** The group $\mathcal{O}_3$ of real rotations in 3 dimensions. Every rotation has a rotation axis $a$ and a rotation angle $\varphi$. By the choice of a unit vector $\vec{a}$ the axis can be directed, and by the corkscrew-rule one direction of rotation can be defined as positive. Now the vector $\varphi\vec{a}$ defines the rotation uniquely, for if $\vec{a}$ is replaced by $-\vec{a}$ and $\varphi$ by $-\varphi$, the rotation remains the same. Hence every rotation $T(s)$ is uniquely determined by the coordinates $s_1$, $s_2$, $s_3$ of the vector $\varphi\vec{a}$. Thus $\mathcal{O}_3$ is seen to be a 3-dimensional Lie Group.

Quite generally, let $U$ be a neighbourhood of the unity element $I$ of a Lie Group $\mathcal{G}$, and let $T(s)$ be the group element in $U$ corresponding to the parameter values $s_1, ..., s_n$. If $T(s)$ and $T(t)$ are sufficiently near to unity, $T(s) T(t)$ and $T(s)^{-1}$ will again lie in $U$, hence we can write

$$T(s) \, T(t) = T(u) \,,$$

$$T(s)^{-1} = T(v) \,.$$

The parameters $u$ determining the product $T(s) T(t)$ are supposed to be continuous functions of the $s$ and $t$, and the $v$ to be continuous functions of the $s$:

$$u = f(s, t) ; \quad v = g(s) .$$

According to a theorem of Gleason, Montgomery, and Zippin[1], the parametrization can always be chosen in such a way that $f$ and $g$ are analytic functions, i.e. power series. In the case of a one-dimensional Lie Group it is even possible to write the laws of composition as

$$u = s + t ; \quad v = -s .$$

Therefore, from now on $I$ shall assume $f$ and $g$ to be analytic functions, and in the one-dimensional case $I$ shall assume

$$T(s) T(t) = T(s + t) ,$$

$$T(s)^{-1} = T(-s) .$$

## B. One-dimensional Lie Groups and Semi-Groups

The group of translations along the $t$-axis is isomorphic to the additive group of real numbers. How can we find all representations of this group by linear transformations of a vector space $\mathscr{V}$ into itself? More precisely: How can we find linear transformations $T(s)$ of $\mathscr{V}$ into itself which are continuous functions of $s$ satisfying the conditions

(18.1) $$T(s) T(t) = T(s + t) ,$$

(18.2) $$T(0) = I?$$

If $\mathscr{V}$ is finite-dimensional and if the matrix-valued function $T(s)$ is supposed to have a continuous derivative, the solution is easy. Differentiating (18.1) with respect to $s$ and putting $s = 0$, one obtains

(18.3) $$A T(t) = \frac{d}{dt} T(t)$$

---

[1] A. M. Gleason: Groups without small subgroups. Annals of Math. **56**, p. 193. D. Montgomery and L. Zippin: Small Subgroups of Finitedimensional Groups. Ann. of Math. **56**, p. 213 (1952).

where $A$ is the derivative of $T(s)$ at $s = 0$. The solution of this differential equation with the initial condition $T(0) = I$ is

$$T(t) = e^{tA}$$
$$= 1 + tA + \frac{1}{2!}(tA)^2 + \cdots .$$

This power series converges for all $t$. The linear transformation $A$ is called the *infinitesimal transformation* generating the one-dimensional Lie Group. If $A$ is a diagonal matrix having diagonal elements $a_1, a_2, \ldots$ the matrix $e^{tA}$ has diagonal elements $\exp(ta_1), \exp(ta_2), \ldots$ .

If the transformations $T(t)$ are applied to a fixed vector $x_0$, the point

$$x(t) = T(t)\, x_0$$

describes an orbit. From (18.3) we have

$$\frac{d}{dt}\, x(t) = A x(t) ,$$

i.e. the velocity at any point $x$ of the orbit is just $Ax$. Thus, the geometrical picture of an infinitesimal transformation $A$ is just the velocity field of a stationary flow. The transformation $T(t)$ is obtained by following the flow during a certain time $t$.

Elie Cartan and John von Neumann showed that the assumption of differentiability of the function $T(t)$ is not necessary: it is a consequence of (18.1) and the assumption of continuity. For a simple proof see van der Waerden[2]. Hence we have the theorem:

*Every continuous representation of the one-dimensional translation group by linear transformations of a finite-dimensional vector space $\mathscr{V}$ is generated by an infinitesimal linear transformation $A$ and given by*

(18.4)                                  $$T(t) = e^{tA}$$

This theorem was generalized to Hilbert spaces by M. H. Stone[3] and to more general vector spaces by E. Hille and K. Yosida[4]. Stone proved:

*Suppose the unitary transformation $T(t)$ of a Hilbert space $\mathscr{V}$ is a continuous function of the real variable $t$ satisfying the conditions*

$$T(s)\, T(t) = T(s + t) ,$$

$$T(0) = I .$$

[2] B. L. van der Waerden: Stetigkeitssätze für halbeinfache Liesche Gruppen. Math. Zeitschrift **36**, p. 781, footnote 2 (1933).
[3] M. H. Stone: On one-parameter unitary groups in Hilbert space. Ann. of Math. **33**, p. 643 (1932).
[4] See: K. Yosida: Functional Analysis, Chapter IX (2nd ed.). Berlin-Heidelberg-New York: Springer 1968.

*Then a self-adjoint linear operator A exists, defined by*

$$(18.5) \qquad\qquad Ax = i \lim_{s \to 0} \frac{U(s) - I}{s} x$$

*such that*

$$(18.6) \qquad\qquad U(t) = \exp(-itA).$$

Independent of each other, Hille and Yosida proved a still more general theorem. They considered semigroups of bounded linear transformations $T(t)$ of a locally convex vector space $\mathscr{V}$ into itself, where $T(t)$ is a continuous function of $t$, defined for $t \geq 0$ and satisfying the conditions (18.1) and (18.2). Under these conditions they first proved the existence of an infinitesimal operator $A$ defined on an everywhere dense subset of the space $\mathscr{V}$ according to (18.5). If the real part of the complex variable $u$ is sufficiently large, the operator $uI - A$ possesses a bounded inverse

$$(18.7) \qquad\qquad R(u) = (uI - A)^{-1},$$

which is the Laplace transform of $T(s)$ in the following sense:

$$(18.8) \qquad\qquad R(u)x = \int_0^\infty e^{-us} T(s)x \, ds$$

It is well-known that a continuous function $f(t)$ is uniquely determined by its Laplace transform $F(u)$. One can calculate $f(t)$ by means of the well-known Inversion Formula [5]

$$(18.9) \qquad\qquad f(t) = \lim_{M \to \infty} \frac{1}{2\pi i} \int_{\varepsilon - iM}^{\varepsilon + iM} e^{ut} F(u) \, du.$$

If $f(t)$ is bounded, this formula holds for every positive $\varepsilon$. It also holds if the values of $f(t)$ are vectors in Hilbert space; the proof is just the same as in the classical case of a real or complex-valued function $f(t)$. Hence (18.8) can be solved for $T(t)x$ by means of the Inversion Formula

$$(18.10) \qquad\qquad T(t)x = \lim_{M \to \infty} \frac{1}{2\pi i} \int_{\varepsilon - iM}^{\varepsilon + iM} e^{ut} R(u)x \, du$$

Another expression for $T(t)$ was given by Hille and Yosida. If $T(t)$ is unitary and hence $A$ self-adjoint, Yosida's formula reduces to Stone's formula (18.6).

---

[5] The limit is a strong limit in the sense of Hilbert Space topology.

## C. Causality and Translations in Time

Consider a quantum-mechanical system, e.g. an atom or molecule or quantum field, which is isolated from the rest of the world during a certain time $T$. Let the state of the system during a short time from $0 - \varepsilon$ to $0 + \varepsilon$ be given by an element $\Phi_0$ of a Hilbert space. The Principle of Causality requires that the state of the system at any later time $t$, or rather between $t - \varepsilon$ and $t + \varepsilon$, is uniquely determined by $\Phi_0$ according to a law. It seems reasonable to assume that this law can be expressed by a linear transformation $U(t)$:

$$\Phi(t) = U(t)\, \Phi_0 .$$

Finally, we may assume that the transformation $U(t)$ is unitary. In all existing theories, this postulate is always fulfilled.

An immediate consequence of these assumptions is

$$U(s + t) = U(s)\, U(t)$$
$$U(0) = I .$$

It also seems reasonable to assume that in a very small time the state vector $\Phi(t)$ does not change much, which means that $U(t)\,\Phi_0$ is a continuous function of $t$. *Hence we suppose that the unitary transformations $U(t)$ form a continuous representation of the semi-group of non-negative translations in time.*

To this representation we may apply the theorems of Stone, Hille, and Yosida, and conclude that a self-adjoint linear operator exists, which was formerly denoted by $A$ and which we shall now call $\hbar^{-1} H$, such that

(18.11)                             $U(t) = \exp(-i\hbar^{-1} Ht) .$

The operator $-i\hbar^{-1} H$ was defined as derivative of $U(s)$ at $s = 0$. Now if we differentiate the equation

$$U(s + t)\, \Phi_0 = U(s)\, U(t)\, \Phi_0$$

with respect to $s$ and put $s = 0$, we obtain

(18.12)                             $\dfrac{d}{dt}\, \Phi(t) = -i\hbar^{-1}\, H\Phi(t) .$

More precisely: if $\Phi_0$ is any state vector for which $H\Phi_0$ is defined, $H$ can also be applied to $\Phi(t)$, and the differential equation (18.12) holds. It has just the form of the Schrödinger equation, hence we may call $H$ the Hamiltonian or Energy Operator, and we have the theorem:

*If a Law of Causality holds, according to the assumptions stated before, a self-adjoint Hamiltonian H exists such that (18.11) and (18.12) hold.*

The theorems of Hille and Yosida are not only interesting from a theoretical point of view. They also give us a tool for the practical solution of the Schrödinger equation (18.12) by means of a Laplace Transformation. To explain this, let us re-write the equations (18.7) and (18.8) of Hille and Yosida as

$$(18.13) \qquad R(u)\,\Phi_0 = \int_0^\infty e^{-ut}\,\Phi(t)\,dt$$

and

$$(18.14) \qquad R(u)\,\Phi_0 = (uI + i\hbar^{-1}\,H)^{-1}\,\Phi_0\,.$$

The first formula says: $R(u)\,\Phi_0$ is just the Laplace transform of the function $\Phi(t)$. The second formula says that this Laplace transform can be computed from the equation

$$(18.15) \qquad (uI + i\hbar^{-1}\,H)\,R(u)\,\Phi_0 = \Phi_0\,.$$

The formulae (18.14) and (18.15) imply that $R(u)\,\Phi_0$ always belongs to the domain of definition of the operator $H$, no matter how $\Phi_0$ may be chosen in Hilbert space, whereas the Schrödinger equation (18.12) has a meaning only if $\Phi_0$ belongs to the domain of definition of $H$. Hence, one can always compute $\Phi(t)$ by means of its Laplace transform $R(u)\,\Phi_0$, no matter what $\Phi_0$ is. To compute $\Phi(t)$ one can use, once more, the inversion formula

$$(18.16) \qquad \Phi(t) = \lim_{M \to \infty} \frac{1}{2\pi i} \int_{\varepsilon - iM}^{\varepsilon + iM} e^{+ut}\,R(u)\,\Phi_0\,du\,.$$

Interesting applications of the Inversion Formula to Quantum Field Theory were given by K. Ingólfsson and others[6].


## D. The Lie Algebra of a Lie Group[7]

From now on, we shall consider only linear transformations in finite-dimensional vector spaces. Let us first recall that for every such transformation $A$ the series

$$e^{tA} = 1 + tA + \frac{1}{2!}\,(tA)^2 + \cdots$$

[6] K. Ingólfsson: Zur Formulierung der mathematischen Theorie der natürlichen Linienbreite. Helvetica Physica Acta **40**, p. 237 (1967).

[7] For proofs of the theorems used here I refer, once and for all, to the book of H. Freudenthal and H. de Vries: Linear Lie Groups. London: Academic Press 1969.

converges, that the transformations $e^{tA}$ form a one-dimensional Lie Group, and that every one-dimensional Lie Group can thus be obtained from an infinitesimal transformation $A$.

Quite generally, if an $n$-dimensional Lie Group $\mathscr{G}$ consists of linear transformations $T$, the parameters $s_1, \ldots, s_n$ in a neighbourhood of $I$ can be chosen in such a way that the transformations $T(s)$ are power series in $s_1, \ldots, s_n$:

(18.17) $$T(s) = I + s_1 I_1 + \cdots + s_n I_n + \cdots .$$

Differentiating (18.17) with respect to $s_1, \ldots, s_n$ and putting $s = 0$, one obtains $n$ infinitesimal transformations $I_1, \ldots, I_n$. Now all group elements in a neighbourhood $U$ of $I$ can be written as

(18.18) $$T = \exp(t_1 I_1) \exp(t_2 I_2) \ldots \exp(t_n I_n)$$

the $t_1, \ldots, t_n$ being a new set of parameters.

The linear transformations

(18.19) $$J = u_1 I_1 + \cdots + u_n I_n$$

obtained by differentiating (18.17) in the direction of an arbitrary vector $u$, are called the *infinitesimal transformations* of the group $\mathscr{G}$. Every group element in a neighbourhood of $I$ can be written as

$$T = \exp(u_1 I_1 + \cdots + u_n I_n) = \exp(J)$$

and hence belongs to a one-dimensional subgroup generated by an infinitesimal transformation $J$ of $\mathscr{G}$. The $u$ are called *canonical parameters*.

The infinitesimal transformations $J$ defined by (18.19) form an Algebra, the *Lie Algebra* belonging to the group $\mathscr{G}$. In this algebra, three operations are defined:

1. addition
2. multiplication by real numbers
3. the operation $[A, B] = AB - BA$.

The commutator $[A, B]$ is obtained by developing

$$e^{sA} e^{tB} e^{-sA} e^{-tB}$$

in a power series and retaining the term with $st$ only. It follows that the brackets $[A, B]$ always belong to the Lie Algebra. In particular we have

$$[I_i, I_j] = \Sigma \, c_{ij}^k I_k .$$

The $c_{ij}^k$ are called the *structural constants* of the Lie Algebra. They determine the structure of the group in a neighbourhood $U$ of $I$. Two Lie Groups having the same structural constants are *locally isomorphic*. Globally they need not be isomorphic.

**Example.** The infinitesimal transformations $A_1, A_2, A_3$ of the three-dimensional rotation group $\mathcal{O}_3$ are just the infinitesimal rotations $I_1, I_2, I_3$ about the $x$-, $y$-, and $z$-axis. Their matrices are

$$I_1 = \begin{pmatrix} 0 & 0 & 0 \\ 0 & 0 & -1 \\ 0 & 1 & 0 \end{pmatrix}, \quad I_2 = \begin{pmatrix} 0 & 0 & 1 \\ 0 & 0 & 0 \\ -1 & 0 & 0 \end{pmatrix}, \quad I_3 = \begin{pmatrix} 0 & -1 & 0 \\ 1 & 0 & 0 \\ 0 & 0 & 0 \end{pmatrix}.$$

The commutator relations are

(18.20)
$$\begin{cases} I_1 I_2 - I_2 I_1 = I_3 \\ I_2 I_3 - I_3 I_2 = I_1 \\ I_3 I_1 - I_1 I_3 = I_2 . \end{cases}$$

Later on we shall see that the group $\mathcal{O}_3$ is locally isomorphic to the group SU(2) of unitary transformations in 2 dimensions with determinant 1.

### E. Representations of Lie Groups

As in Chapter II, we shall denote the elements of a group $\mathcal{G}$ by letters like $a, b, \ldots$ and the representing matrices by the corresponding capital letters $A, B, \ldots$. We consider only continuous representations, which means that the representing matrix $B$ is supposed to be a continuous function of the group element $b$, defined in a neighbourhood $U$ of the unity element $e$. In such a neighbourhood $U$ the elements $b$ may be represented as products (18.18):

(18.21) $$b(t) = \exp(t_1 I_1) \ldots \exp(t_n I_n).$$

Any continuous representation $b \to B$ yields a continuous representation of the one-dimensional subgroup formed by the elements

(18.22) $$b_1(t_1) = \exp(t_1 I_1).$$

Now we have seen that any such representation is generated by an infinitesimal transformation $A_1$, and that the element (18.22) is represented by the matrix

$$B_1(t_1) = \exp(t_1 A_1).$$

The same holds for $b_2(t_2) = \exp(t_2 I_2)$, and so on. Hence the product (18.21) is represented by

$$(18.23) \qquad\qquad B(t) = \exp(t_1 A_1) \ldots \exp(t_n A_n).$$

This means: the matrix $B(t)$ is an analytic function of $t_1, \ldots, t_n$, and the representation is completely determined by its infinitesimal transformations $A_1, \ldots, A_n$.

The group element

$$\exp(s\, I_1) \exp(t\, I_2) \exp(-s\, I_1) \exp(-t\, I_2)$$

must be represented by the matrix

$$\exp(s\, A_1) \exp(t\, A_2) \exp(-s\, A_1) \exp(-t\, A_2).$$

Expanding in power series and comparing the coefficients of $st$, one sees that to $[I_1, I_2]$ corresponds $[A_1, A_2]$, and just so for any commutator $[I_i, I_j]$. Thus we see that the equations

$$[I_i, I_j] = \Sigma\, c_{ij}^k\, I_k$$

which hold for the $I_i$, must hold just as well for the $A_i$:

$$(18.24) \qquad\qquad [A_i, A_j] = \Sigma\, c_{ij}^k\, A_k.$$

Thus, the problem to find all continuous representations of $\mathscr{G}$ can be reduced to a much simpler algebraic problem: To find a set of $n$ matrices $A_1, \ldots, A_n$ satisfying the equations (18.24).

Any solution of this algebraic problem leads to a local representation of $\mathscr{G}$, i.e. to a representation of a neighbourhood of $e$ in $\mathscr{G}$, in which products $bc$ are represented by products $BC$. If the representation is extended to the whole group, it is possible that one obtains a multi-valued representation.

## § 19. The Unitary Groups SU(2) and the Rotation Group $\mathscr{O}_3$

Let me introduce the following notations:

SL(2) (SL = Special Linear) means: Group of all linear transformations of 2 complex variables with determinant 1.

SU(2) (SU = Special Unitary) means: Group of all unitary transformations of 2 complex variables with determinant 1.

$\mathcal{O}_3$ ($\mathcal{O} =$ Orthogonal) means: Group of rotations of a real 3-dimensional vector space.

Consider the complex vector space formed by the linear forms $c_1 u_1 + c_2 u_2$ in two indeterminates $u_1, u_2$ with complex coefficients $c_1, c_2$. The transformations of the Special Linear Group SL(2) transform the basic vectors $u_1, u_2$ of this vector space into

$$u_1' = u_1 \alpha + u_2 \gamma,$$
$$u_2' = u_1 \beta + u_2 \delta.$$

The corresponding transformations of the coefficients $c_1, c_2$ of a vector $c_1 u_1 + c_2 u_2$ are

$$c_1' = \alpha c_1 + \beta c_2,$$
$$c_2' = \gamma c_1 + \delta c_2,$$

and the matrices of these transformations are

$$A = \begin{pmatrix} \alpha & \beta \\ \gamma & \delta \end{pmatrix}, \quad \alpha\delta - \beta\gamma = 1.$$

By direct verification one sees that the inverse matrix of $A$ is

$$A^{-1} = \begin{pmatrix} \delta & -\beta \\ -\gamma & \alpha \end{pmatrix}$$

Unitary transformations $A$ are those which leave invariant a positive Hermitian form; they form the *special unitary group* SU(2).

We may always assume that the Hermitian form is the unit form

$$c_1^* c_1 + c_2^* c_2.$$

In this case, the condition for $A$ to be unitary is—see (9.8)—

or
$$A^{-1} = A^\dagger$$

$$\begin{pmatrix} \delta & -\beta \\ -\gamma & \alpha \end{pmatrix} = \begin{pmatrix} \alpha^* & \gamma^* \\ \beta^* & \delta^* \end{pmatrix}$$

which implies $\alpha^* = \delta$ and $\beta^* = -\gamma$. Hence SU(2) consists of all matrices

(19.1) $$\begin{pmatrix} \alpha & \beta \\ -\beta^* & \alpha^* \end{pmatrix} \quad \text{with} \quad \alpha\alpha^* + \beta\beta^* = 1.$$

From (19.1) one sees that a unitary transformation transforms the pair $(c_1, c_2)$ just as it transforms the pair $(-c_2^*, c_1^*)$. We shall need this result presently.

The vector space of linear forms $c_1 u_1 + c_2 u_2$ can be imbedded in the polynomial domain $\mathbb{C}[u_1, u_2]$ consisting of all polynomials in $u_1$ and $u_2$ with complex coefficients. Now a representation of the groups SL(2) and SU(2) can be obtained as follows. Consider the vector space of all forms of degree $k$

$$c_0 u_1^k + c_1 u_1^{k-1} u_2 + \cdots + c_k u_2^k .$$

The basis vectors of this vector space are the $k+1$ power-products

$$(19.2) \qquad\qquad w_r = u_1^r u_2^{k-r} \quad (r = 0, 1, \ldots, v) .$$

The transformation $A$ transforms such a power product into

$$u_1^{\prime r} u_2^{\prime k-r} = (u_1 \alpha + u_2 \gamma)^r (u_1 \beta + u_2 \delta)^{k-r} ,$$

which is a linear combination of the power products (19.2). The representation of SL(2) or SU(2) thus obtained will be denoted by $\varrho_J$, where $J$ is defined by

$$J = \tfrac{1}{2} k .$$

In particular, $\varrho_0$ is the identity representation $A \to 1$ of degree 1, and $\varrho_{\frac{1}{2}}$ is the representation of $A$ by $A$. The representation $\varrho_J$ has degree

$$k + 1 = 2J + 1 .$$

In §20 we shall prove that the representations $\varrho_J$ are irreducible.

The representation $\varrho_1$ in the space of quadratic forms

$$(19.3) \qquad\qquad c_0 u_1^2 + c_1 u_1 u_2 + c_2 u_2^2$$

leaves invariant the discriminant of the quadratic form

$$c_1^2 - 4 c_0 c_2 .$$

Introducing instead of $c_0, c_1, c_2$ new variables by means of substitution

$$(19.4) \qquad \begin{cases} x = -c_0 + c_2 \\ y = -i(c_0 + c_2) \\ z = c_1 \end{cases} \quad \begin{cases} x + iy = 2c_2 \\ x - iy = -2c_0 \\ z = c_1 \end{cases}$$

we find

$$x^2 + y^2 + z^2 = (x + iy)(x - iy) + z^2 = c_1^2 - 4c_0c_2 \, .$$

Hence, the transformations occuring in the representation $\varrho_1$ leave the form $x^2 + y^2 + z^2$ invariant: they are *orthogonal transformations*. Since their determinant is $+1$, they are *rotations* of 3-space.

Are these rotations real? The coefficients $c_0$, $c_1$, $c_2$ of an arbitrary quadratic form are transformed in exactly the same way as the coefficients $a_1 b_1$, $a_1 b_2 + a_2 b_1$, $a_2 b_2$ of the special form $(a_1 u_1 + a_2 u_2)(b_1 u_1 + b_2 u_2)$. Under a unitary transformation (19.1) $b_1, b_2$ are transformed exactly as $-a_2^*, a_1^*$. Consequently, the $c_j$ are transformed like

$$-a_1 a_2^*, \; a_1 a_1^* - a_2 a_2^*, \, a_2 a_1^* \, .$$

Hence, the $x, y, z$ defined by (19.4) are transformed like

$$a_1 a_2^* + a_2 a_1^*; \quad i(a_1 a_2^* - a_2 a_1^*); \quad a_1 a_1^* - a_2 a_2^* \, .$$

Now, these three numbers are real and are always transformed into real numbers, so that the transformation coefficients must also be real. That is:

*The vectors $(x, y, z)$ always suffer real rotations under the representation $\varrho_1$ of the group* SU(2).

It is easy to see that every real rotation of 3-space actually occurs in the representation $\varrho_1$. To show this it is sufficient to calculate the rotations corresponding to the special unitary transformations

$$B(\beta) = \begin{pmatrix} \cos\beta & -\sin\beta \\ \sin\beta & \cos\beta \end{pmatrix}, \quad C(\gamma) = \begin{pmatrix} e^{-i\gamma} & 0 \\ 0 & e^{+i\gamma} \end{pmatrix}.$$

One finds:

(19.5)

$$B(\beta): \begin{cases} x' = & x\cos 2\beta + z\sin 2\beta \\ y' = & y \\ z' = & -x\sin 2\beta + z\cos 2\beta \end{cases}$$

$$C(\gamma): \begin{cases} x' = x\cos 2\beta - y\sin 2\beta \\ y' = x\sin 2\beta + y\cos 2\beta \\ z' = z \, . \end{cases}$$

Hence $B(\beta)$ and $C(\gamma)$ are rotations about the $y$- and $z$-axis with rotation angles $2\beta$ and $2\gamma$. Now every rotation can be obtained as a product of such special rotations. In fact, a rotation $R$ with Euler angles $\vartheta$, $\varphi$, $\psi$ is just the product $Z_\varphi\, Y_\vartheta\, Z_\psi$ of rotations about the $z$-, $y$- and $z$-axis with angles of rotation $\varphi$, $\vartheta$, $\psi$. Hence, to obtain the rotation $R$ in the representation $\varrho_1$ we need only form the matrix

$$A = C(\tfrac{1}{2}\varphi)\, B(\tfrac{1}{2}\vartheta)\, C(\tfrac{1}{2}\psi)$$

$$
(19.6) \quad = \begin{pmatrix} e^{-\frac{1}{2}i\varphi} & 0 \\ 0 & e^{+\frac{1}{2}i\varphi} \end{pmatrix} \begin{pmatrix} \cos\tfrac{1}{2}\vartheta & -\sin\tfrac{1}{2}\vartheta \\ \sin\tfrac{1}{2}\vartheta & \cos\tfrac{1}{2}\vartheta \end{pmatrix} \begin{pmatrix} e^{-\frac{1}{2}i\psi} & 0 \\ 0 & e^{+\frac{1}{2}i\psi} \end{pmatrix}
$$

$$
= \begin{pmatrix} e^{-\frac{1}{2}i(\varphi+\psi)}\cos\tfrac{1}{2}\vartheta & -e^{-\frac{1}{2}i(\varphi-\psi)}\sin\tfrac{1}{2}\vartheta \\ e^{+\frac{1}{2}i(\varphi-\psi)}\sin\tfrac{1}{2}\vartheta & e^{+\frac{1}{2}i(\varphi+\psi)}\cos\tfrac{1}{2}\vartheta \end{pmatrix} = \begin{pmatrix} \alpha & \beta \\ -\beta^* & \alpha^* \end{pmatrix}
$$

In order to investigate the fidelity of the representation $\varrho_1$ it is sufficient, according to the Homomorphy Theorem (§ 10), to find out which transformations of SU(2) yield the identity in the representation $\varrho_1$. These transformations must leave the products $u_1^2$, $u_1 u_2$ and $u_2^2$ invariant and this is obviously done only by the two transformations

$$I = \begin{pmatrix} 1 & 0 \\ 0 & 1 \end{pmatrix} \quad \text{and} \quad -I = \begin{pmatrix} -1 & 0 \\ 0 & -1 \end{pmatrix}.$$

The Homomorphism Theorem now says that there is a one-to-one correspondence between the pairs of matrices $(A, -A)$ in SU(2) and the rotations $R$ in $\mathcal{O}_3$:

$$R \rightleftarrows (A, -A)$$

and that this correspondence is an isomorphism. If one restricts oneself to a neighbourhood of unity, the correspondence $R \rightleftarrows A$ is one-to-one. These facts are expressed by saying that SU(2) is a *two-valued representation of* $\mathcal{O}_3$, and that $\mathcal{O}_3$ is *locally isomorphic* to SU(2).

The representations $\varrho_J$ were defined as representations of SU(2), and SU(2) is a two-valued representation of $\mathcal{O}_3$, hence the $\varrho_J$ can be regarded as at most two-valued representations of $\mathcal{O}_3$. More precisely: to every rotation $R$ corresponds a pair $(A, -A)$ in SU(2), and to this pair corresponds a pair

$$(\varrho_J A, \varrho_J(-A))$$

in the representation $\varrho_J$ of SU(2). If $J$ is integer, $k = 2J$ is even, and the two matrices $\varrho_J A$ and $\varrho_J(-A)$ coincide, but if $k$ is odd, they do not coincide. Hence $\varrho_J$ is a one-valued representation of $\mathcal{O}_3$ if $2J$ is even, and a two-valued representation of $\mathcal{O}_3$ if $2J$ is odd.

We now prove:

*The representation $\varrho_J$ of $SU(2)$ possesses an invariant Hermitian form, namely*

(19.7)
$$\sum_{r=0}^{k} r!\,(k-r)!\,c_r^*\,c_r\,.$$

*Proof.* The coefficients $c_r$ are transformed just as the coefficients of the special form $(a_1 u_1 + a_2 u_2)^k$, i.e. just as

$$d_r = \binom{k}{r}\,a_1^{k-r}\,a_2^r\,.$$

Likewise, the $c_r^*$ are transformed as

$$d_r^* = \binom{k}{r}\,a_1^{*\,k-r}\,a_2^{*\,r}\,.$$

Since $a_1^* a_1 + a_2^* a_2$ remains invariant, so does

$$k!\,(a_1^* a_1 + a_2^* a_2)^k = k!\,\sum_{r=0}^{k} \binom{k}{r}\,(a_1^*)^{k-r}\,a_1^{k-r}\,(a_2^*)^r\,a_2^r$$

$$= \Sigma\, r!\,(k-r)!\,d_r^*\,d_r\,.$$

Hence, (19.7) remains invariant as well.

Thus, the representations $\varrho_J$ of $SU(2)$ are *unitary*. The vectors

(19.8)
$$\frac{u_1^{k-r} u_2^r}{\sqrt{r!\,(k-r)!}}$$

form a normed orthogonal system for the form (19.7).

To the rotation $R = Z_\gamma$ about the $z$-axis with rotation angle $\gamma$ corresponds, as we have seen before, a matrix

$$C(\tfrac{1}{2}\gamma) = \begin{pmatrix} e^{-\frac{1}{2}i\gamma} & 0 \\ 0 & e^{\frac{1}{2}i\gamma} \end{pmatrix}.$$

The transformation $C(\tfrac{1}{2}\gamma)$ multiplies the basis vector $u_1$ by $e^{-\frac{1}{2}i\gamma}$ and $u_2$ by $e^{\frac{1}{2}i\gamma}$. In the representation $\varrho_J$ this transformation is represented by a diagonal matrix: the basis vector

$$u_1^{k-r} u_2^r$$

is multiplied by

$$e^{-\frac{1}{2}i\gamma(k-r)+\frac{1}{2}i\gamma r} = e^{(-J+r)i\gamma}\,.$$

Since $r$ goes from 0 to $k = 2J$, these diagonal elements are

$$e^{-Ji\gamma}, e^{(-J+1)i\gamma}, \ldots, e^{+Ji\gamma}.$$

To the rotation $Y_\pi$ about the $y$-axis with angle $\pi$ corresponds, according to (19.5), the matrix

$$B\left(\frac{\pi}{2}\right) = \begin{pmatrix} 0 & -1 \\ 1 & 0 \end{pmatrix}$$

which transforms $u_1$ into $u_2$ and $u_2$ into $-u_1$. In the representation $\varrho_J$ one obtains a transformation which takes

$$u_1^{k-r} u_2^r$$

into

$$(-1)^r u_1^r u_2^{k-r}.$$

## § 20. Representations of the Rotation Group $\mathcal{O}_3$

As we have seen, the infinitesimal transformations of the threedimensional Rotation Group satisfy the equations (18.18). In any representation, the infinitesimal transformations $A_1, A_2, A_3$ corresponding to $I_1, I_2, I_3$ have to satisfy the same equations

(20.1)
$$\begin{cases} A_1 A_2 - A_2 A_1 = A_3 \\ A_2 A_3 - A_3 A_2 = A_1 \\ A_3 A_1 - A_1 A_3 = A_2. \end{cases}$$

The same relations also hold for multi-valued representations.

As in § 8, we introduce the operators

$$L_x = i A_1, \quad L_y = i A_2, \quad L_z = i A_z$$

and we put

$$L_x + i L_y = L_p,$$
$$L_x - i L_y = L_q.$$

The relations (20.1) can now be written as

(20.2)
$$\begin{cases} L_z L_p - L_p L_z = L_p \\ L_z L_q - L_q L_z = -L_q \\ L_p L_q - L_q L_p = 2L_z. \end{cases}$$

Assume that an arbitrary representation of the rotation group in a (finite-dimensional) vector space $\mathscr{V}$ is given. It induces a representation of the Abelian subgroup consisting of the rotations $(0, 0, \gamma)$ about the $z$-axis. This representation can be decomposed according to § 12, Example 2, and one finds a set of basis vectors $v_M$ which will be multiplied by $e^{-iM\gamma}$ under the rotation $Z_\gamma = (0, 0, \gamma)$. If the representation is univalued, $M$ must be integral; but this need not be the case if the representation is uni-valued only in a neighborhood of unity. We have

$$L_z v_M = i I_z v_M = i\left(\frac{\partial}{\partial \gamma} Z_\gamma v_M\right)_{\gamma=0} = i\left(\frac{\partial}{\partial \gamma} e^{-iM\gamma} v_M\right)_{\gamma=0} = M v_M$$

so that the $v_M$ are eigenvectors of the operator $L_z$ with eigenvalues $M$.

**Lemma.** *If the vector $v$ corresponds to the eigenvalue $M$ of $L_z$, then $L_p v$ corresponds to the eigenvalue $(M + 1)$ and $L_q v$ to the eigenvalue $(M - 1)$ of $L_z$.*

*Proof.* From $L_z v = M v$ it follows that

$$L_z L_p v = (L_p L_z + L_p) v = L_p M v + L_p v = (M + 1) L_p v$$

and correspondingly

$$L_z L_q v = (M - 1) L_q v .$$

Thus, the lemma is proved.

We now look for a vector $v_J$ in the space $\mathscr{V}$ corresponding to the largest occuring eigenvalue $J$ of $L_z$ (or, if imaginary eigenvalues occur, corresponding to one with largest real part). Then $L_p v_J$ corresponds to the eigenvalue $J + 1$. Since, however, $J$ has the largest possible real part, $L_p v_J$ must be zero. Furthermore,

$$v_{J-1} = L_q v_J \text{ corresponds to the eigenvalue } J - 1,$$

$$v_{J-2} = L_q v_{J-1} \text{ corresponds to the eigenvalue } J - 2,$$

etc. The series is continued until one encounters the zero vector, which must occur sooner or later, since only finitely many eigenvectors can occur in the space $\mathscr{V}$.

It is now easy to prove for $M = J, J - 1, J - 2, \ldots$ that

(20.3) $$L_p v_M = c_M v_{M+1}; \qquad c_M = \text{integer} .$$

Indeed, the formula is correct for the largest value $M = J$ with $c_J = 0$ since $L_p v_J = 0$. We now show that (20.3) is correct for $M = \mu - 1$ if it is correct for $M = \mu$. In fact,

$$L_p v_{\mu - 1} = L_p L_q v_\mu = (L_q L_p + 2 L_z) v_\mu = L_q c_\mu v_{\mu + 1} + 2 \mu v_\mu$$
$$= (c_\mu + 2 \mu) v_\mu$$

which proves (20.3). For the $c_M$ we have the recursion formula

$$c_{\mu - 1} = c_\mu + 2 \mu; \quad c_J = 0,$$

with solution

(20.4) $$c_M = J(J + 1) - M(M + 1).$$

It must sooner or later happen that $v_M = 0$ while the preceding $v_{M+1}$ is not zero. Then $c_M$ must be zero. But then $M = -(J + 1)$, since the equation

$$J(J + 1) - x(x + 1) = 0$$

has only the roots $x = J$ and $x = -(J + 1)$, and the value $M = J$ does not come into consideration because of $v_J \neq 0$. Hence, the first zero vector in the sequence of vectors $v_J, v_{J-1}, v_{J-2}, \dots$ is $v_{-(J+1)}$. The number of terms in the sequence $v_J, v_{J-1}, \dots, v_{-J}$ is $2J + 1$, so that $2J + 1$ *is an integer, and $J$ is half an integer.* The possible values of $J$ are

$$J = 0, \tfrac{1}{2}, 1, 1\tfrac{1}{2}, \dots.$$

To achieve greater symmetry in the formulae, one can provide the $v_M$ with numerical factors and define

$$L_q v_M = \sqrt{J(J + 1) - M(M - 1)} \cdot v_{M - 1}.$$

Then

(20.5) $$\begin{cases} L_p v_M = \sqrt{J(J + 1) - M(M + 1)} \cdot v_{M+1} \\ \quad = \sqrt{(J - M)(J + M + 1)} \cdot v_{M+1}, \\ L_q v_M = \sqrt{J(J + 1) - M(M - 1)} \cdot v_{M-1} \\ \quad = \sqrt{(J + M)(J - M + 1)} \cdot v_{M-1}, \\ L_z v_M = M v_M. \end{cases}$$

The subspace $(v_J, v_{J-1}, \dots, v_{-J})$ of the vector space $\mathscr{V}$ is transformed into itself by the operations $L_p, L_q, L_z$, hence also by the infinitesimal rotations $A_x, A_y, A_z$. This implies that this subspace is also transformed into itself by all transformations of the representation of the rotation group, i.e.:

*The vectors $v_J, v_{J-1}, \ldots, v_{-J}$ determine an invariant subspace $\mathscr{V}_{2J+1}$.*

The transformations of this subspace form a representation of the rotation group which is completely determined by the equations (20.5). In the space $V_{2J+1}$ the operator $L_z$ has the simple eigenvalues $M = J$, $J - 1, \ldots, -J$. It is also worth noting that all vectors of the space $\mathscr{V}_{2J+1}$ are eigenvectors of the operator

$$\mathscr{L}^2 = L_x^2 + L_y^2 + L_z^2 = \tfrac{1}{2}(L_p L_q + L_q L_p) + L_z^2 .$$

Indeed, from (20.5) we get by a simple calculation:

(20.6) $$\mathscr{L}^2 v_M = J(J+1) v_M .$$

*The space $\mathscr{V}_{2J+1}$ is irreducible.* For if $\mathscr{V}'$ is an invariant subspace of $V_{2J+1}$ and $v'$ is an eigenvector of $L_z$ in the space $\mathscr{V}'$, then $v'$ must coincide up to a factor with one of the vectors $v_J, \ldots, v_{-J}$. Indeed, there are no other eigenvectors of $L_z$ in $\mathscr{V}_{2J+1}$. According to (20.5) the transformations $L_q$ and $L_p$ generate from one $v' = v_{M'}$ all other $v_M (M = J$, $J - 1, \ldots, -J)$. Hence, all of these $v_M$ belong to $\mathscr{V}'$ so that $\mathscr{V}'$ is the whole space $\mathscr{V}_{2J+1}$, which was to be proved.

*The representation of degree $2J + 1$ determined by (20.5) is equivalent to the representation denoted by $\varrho_J$ constructed in the previous section.* In fact, if we apply a rotation $Z_\gamma$ to the basis vectors of the representation space of $\varrho_J$, i.e. to the vectors

$$w_r = u_1^{2J-r} u_2^r \quad (r = 0, 1, \ldots, 2J),$$

they are multiplied, as we have seen, by

$$e^{i\gamma(r-J)}$$

hence the infinitesimal transformation $A_z$ causes the same vectors to be multiplied by $i(r - J)$:

$$A_z w_r = i(r - J) w_r .$$

Hence we have

$$L_z w_r = iA_z w_r = (J - r) w_r ,$$

which means that the $w_r$ are eigenvectors of $L_z$ with eigenvalues

$$J - r = J, J - 1, \ldots, -J .$$

If we now construct the sequence of vectors $v_M$ as before, starting with the vector $v_J = w_0$ belonging to the highest eigenvalue $J$, we see that the space $\mathscr{V}_{2J+1}$ generated by the vectors $v_J, v_{J-1}, \ldots, v_{-J}$ is just the same as the representation space of $\varrho_J$ generated by the vectors $w_0, w_1, \ldots, w_{2J}$. This proves the equivalence of the representations.

The basis vectors $v_M$ of the space $\mathscr{V}_{2J-1}$ must be identical with the products $u_1^{J+M} u_2^{J-M}$ defining the representation $\varrho_J$, multiplied by numerical factors. If one calculates these factors, one finds

$$(20.7) \qquad v_M = \frac{u_1^{J+M} u_2^{J-M}}{\sqrt{(J+M)!\,(J-M)!}}.$$

According to § 19, these $v_M$ constitute a normed orthogonal system.

In the same way one can show, if $J$ is an integer $l$, that the representation $\varrho_l$ defined by (20.5) is the same as the representation defined by the spherical harmonics $Y_l^{(m)}$. In fact, there are just $2l+1$ spherical harmonics $Y_l^{(m)}$ ($m = l, l-1, \ldots, -l$), and the largest eigenvalue of the operator $L_z$ in the space of spherical harmonics of order $l$ is $m = l$. Hence:

The spherical harmonics $Y_l^{(m)}$ are transformed according to the irreducible representation $\varrho_l$. It is now seen that for integer values of $J$ the representation $\varrho_J$ is one-valued. On the other hand, if $J$ is an integer plus $\frac{1}{2}$, the representation $\varrho_J$ is not one-valued, for, if a rotation $(0, 0, \gamma)$ is applied to the vector $v_J$, this vector is multiplied by $\exp(-iJj)$. If $\gamma = 2\pi$ and $J = n + \frac{1}{2}$, this factor becomes $-1$. Hence, the identical rotation 1 is represented in $\varrho_{n+\frac{1}{2}}$ by the matrices 1 and $-1$.

In the special case $J = \frac{1}{2}$, we have just 2 basis vectors

$$v_{\frac{1}{2}} = u_1, v_{-\frac{1}{2}} = u_2,$$

and the group $\mathscr{O}_3$ is represented by the unitary group SU(2). In this special case, the formulae (20.5) yield

$$L_p = L_x + iL_y = \begin{pmatrix} 0 & 1 \\ 0 & 0 \end{pmatrix},$$

$$L_q = L_x - iL_y = \begin{pmatrix} 0 & 0 \\ 1 & 0 \end{pmatrix},$$

$$L_z = \frac{1}{2}\begin{pmatrix} 1 & 0 \\ 0 & -1 \end{pmatrix}$$

and hence

$$L_x = iA_x = \tfrac{1}{2}\begin{pmatrix} 0 & 1 \\ 1 & 0 \end{pmatrix}$$

$$L_y = iA_y = \tfrac{1}{2}\begin{pmatrix} 0 & -i \\ i & 0 \end{pmatrix},$$

$$L_z = iA_z = \tfrac{1}{2}\begin{pmatrix} 1 & 0 \\ 0 & 1 \end{pmatrix}.$$

The three matrices

$$s_1 = \begin{pmatrix} 0 & 1 \\ 1 & 0 \end{pmatrix}, \quad s_2 = \begin{pmatrix} 0 & -i \\ i & 0 \end{pmatrix}, \quad s_3 = \begin{pmatrix} 1 & 0 \\ 0 & 1 \end{pmatrix}$$

are called *Pauli Matrices*; they were introduced by Pauli in the theory of spinning electron.

It is now easy to prove the *Main Theorem*:

*Every irreducible representation $\varrho$ of $\mathcal{O}_3$ is equivalent to one of the representations $\varrho_J$.*

*Proof.* In the vector space $\mathscr{V}$ of the representation one can always construct an invariant subspace $\mathscr{V}_{2J+1}$ by the method previously exposed. If $\mathscr{V}$ is irreducible, the subspace coincides with the whole space $\mathscr{V}$.

H. Weyl has proved that every representation of SU(2) or $\mathcal{O}_3$ is unitary and hence completely reducible[8]. The considerations explained in this section yield a very simple method to find the irreducible representations $\varrho_J$ contained in a given representation $\varrho$. The method is as follows: One makes a list of all eigenvalues of the operator $L_z$ with their multiplicities. The eigenvalues are integers or integers plus $\tfrac{1}{2}$. If $J$ is the largest occurring eigenvalue, then there exists a representation $\varrho_J$ contained in $\varrho$ in whose space the eigenvalues $J, J-1, \ldots, -J$ each occur once. One then seeks the largest eigenvalue $J'$ from those remaining, splits off a representation $\varrho_{J'}$, etc. until all of the eigenvalues are used up.

## § 21. *Examples and Applications*

### A. The Product Representation $\varrho_j \times \varrho_{j'}$

Let the space of the representation $\varrho_j$ be $(U_j, \ldots, U_{-j})$ and let that of $\varrho_{j'}$ be $(V_{j'}, \ldots, V_{-j'})$. Then the space of the representation $\varrho_j \times \varrho_{j'}$ has all products $U_m V_{m'}$ as basis vectors. The product $U_m V_{m'}$ assumes the factor

---

[8] Weyl's three beautiful papers: Theorie der Darstellung kontinuierlicher halb-einfacher Gruppen durch lineare Transformationen in Math. Zeitschrift **23** (1925) and **24** (1926) have been reprinted in "Selecta Hermann Weyl". Basel: Birkhäuser 1956.

$e^{-i(m+m')\gamma}$ under a rotation $(0, 0, \gamma)$ about the $z$-axis and thus corresponds to the eigenvalue $M = m + m'$ of the operator $L_z$. The possible values of $M$ are:

$$(m = j) \qquad M = j + j', j + j' - 1, \ldots\ldots, j - j',$$

$$(m = j - 1) \quad M = \qquad j + j' - 1, j + j' - 2, \ldots, j - j' - 1,$$

$$(m = j - 2) \quad M = \qquad\qquad\qquad j + j' - 2, \ldots\ldots, j - j' - 2.$$

$$\cdots\cdots\cdots\cdots\qquad \cdots\cdots\cdots\cdots\cdots\cdots\cdots\cdots\cdots\cdots\cdots\cdots$$

$$(m = -j) \quad M = \qquad\qquad\qquad\qquad -j + j', \ldots\ldots, \; -j - j'.$$

We can assume, say, $j \geq j'$. One then sees that the largest value $M = j + j'$ occurs *once*, the next largest $M = j + j' - 1$ *twice*, etc., each time with an increase of one until the value $M = j - j'$ is reached, which thus occurs $(2j' + 1)$ times. All values from $M = j - j'$ to $M = -(j + j')$ occur the same number of times.

The rules of § 20 imply the following: the representation $\varrho_{j+j'}$, which corresponds to the largest eigenvalue is contained *once* in the representation $\varrho_j \times \varrho_{j'}$; its representation space makes use of each of the eigenvalues $M = j + j', \ldots, -(j + j')$ once. After these have been deleted, the largest remaining eigenvalue is $M = j + j' - 1$, which now occurs only once, hence, the representation $\varrho_{j+j'-1}$ likewise occurs only once. Proceeding in this way one finally obtains the representation $\varrho_{j-j'}$ which removes all of the remaining eigenvalues. Consequently,

$$(21.1) \qquad\qquad \varrho_j \times \varrho_{j'} = \varrho_{j+j'} + \varrho_{j+j'-1} + \cdots + \varrho_{|j-j'|}.$$

The addition of the absolute value sign to $j - j'$ makes the formula symmetric in $j$ and $j'$, and consequently, it holds also for $j' > j$. For example:

$$\varrho_0 \times \varrho_j = \varrho_j,$$

$$\varrho_1 \times \varrho_1 = \varrho_2 + \varrho_1 + \varrho_0,$$

$$\varrho_1 \times \varrho_{\frac{1}{2}} = \varrho_{1\frac{1}{2}} + \varrho_{\frac{1}{2}}.$$

## B. The Clebsch-Gordan Series

In a first reading, this subsection may be left out.

In order to carry out the reduction of the representation $\varrho_j \times \varrho_{j'}$ *explicitly*, we must actually exhibit vectors $W_M$ in the space of products

$U_m V_{m'}$, which are transformed according to $\varrho_j$, where $J$ is any one of the numbers $j+j'$, $j+j'-1,...,j-j'$. We may set according to (20.7)

$$U_m^{\cdot} = \frac{u_1^{j+m} u_2^{j-m}}{\sqrt{(j+m)!\,(j-m)!}}\;;\quad V_{m'} = \frac{v_1^{j'+m'} v_2^{j'-m'}}{\sqrt{(j'+m')!\,(j'-m')!}}\;.$$

We now form for $J = j+j' - \lambda$ $(\lambda = 0, 1, 2,..., 2j')$ the expression

$$(21.2)\quad P = (u_1 v_2 - u_2 v_1)^\lambda (u_1 x_1 + u_2 x_2)^{2j-\lambda} (v_1 x_1 + v_2 x_2)^{2j'-\lambda}$$

and claim that the coefficients of

$$(21.3)\qquad\qquad X_M^J = \frac{x_1^{J+M} x_2^{J-M}}{\sqrt{(J+M)!\,(J-M)!}}$$

in $P$ represent the desired quantities $W_M^J$ for $M = J, J-1,..., -J$. To show that this claim is justified, we have to prove that the coefficients $W_M^J$ are transformed just like products

$$U_M = \frac{u_1^{J+M} u_2^{J-M}}{\sqrt{(J+M)!\,(J-M)!}}$$

The proof can be given by the methods of the Theory of Invariants. In this theory, the notions *covariant* and *contravariant* play a fundamental role. I shall first explain these notions.

Consider the polynomial domain

$$\mathbb{C}[u_1, u_2; v_1, v_2; x_1, x_2]$$

where $\mathbb{C}$ is the field of complex numbers. In this domain the linear forms $u_1 c_1 + u_2 c_2$ form a two-dimensional vector space. A linear transformation with matrix $A$ transforms the vectors of this space as follows

$$u_k' = \Sigma u_i a_{ik}\,.$$

In matrix notation, this formula can be written as

$$(21.4)\qquad\qquad (u_1', u_2') = (u_1, u_2) \cdot A\,.$$

We shall suppose that the variables $v_1, v_2$ are *covariant* to $u_1, u_2$, *i.e.* that they are transformed just as the $u$:

$$(21.5)\qquad\qquad (v_1', v_2') = (v_1, v_2) A$$

We can unite (21.4) and (21.5) to one single matrix formula:

$$\begin{pmatrix} u_1' & u_2' \\ v_1' & v_2' \end{pmatrix} = \begin{pmatrix} u_1 & u_2 \\ v_1 & v_2 \end{pmatrix} \cdot A \,.$$

This implies

$$u_1' v_2' - u_2' v_1' = u_1 v_2 - u_2 v_1$$

since the determinant of $A$ was always supposed to be 1. Hence: *The determinant*

$$(u, v) = u_1 v_2 - u_2 v_1$$

*is an invariant under all transformations of the special linear group* SL(2).

We now agree to transform the variables $x_1$ and $x_2$ in such a way that the expression

(21.6)                           $$u_x = u_1 x_1 + u_2 x_2$$

is invariant:

(21.7)                           $$u_1' x_1' + u_2' x_2' = u_1 x_1 + u_2 x_2 \,.$$

As we have seen in § 14, the invariance of $u_x$ implies that the matrices of transformation of the variables $x_1, x_2$ are just the inverse transposed of the original matrices $A$. Variables $x_1, x_2$ which are transformed in this way are called *contravariant* variables.

It is easy to generalize these notions to $n$ variables. The variables $x_1, \ldots, x_n$ are contravariant to $u_1, \ldots, u_n$, if the sum $u_1 x_1 + \cdots + u_n x_n$ remains invariant. In the "Ricci Calculus" as used by Einstein in General Relativity, covariant and contravariant variables are distinguished by lower and upper indices, and the sum $\sum u_k x^k$ is an invariant.

The product $P$ defined by (21.2), being a product of invariants, is invariant. If we write this product as

$$P = \sum_M W_M^J X_M^J \,,$$

where the $X$ are defined by (21.3), it follows that the $W$ are transformed contravariant to the $X$. On the other hand, the expression

$$(u_1 x_1 + u_2 x_2)^{2J} = \sum_M \binom{2J}{M} u_1^{J+M} u_2^{J-M} x_1^{J+M} x_2^{J-M}$$

$$= \sum_M (2J)! \, \frac{u_1^{J+M} u_2^{J-M}}{\sqrt{(J+M)!\,(J-M)!}} \, \frac{x_1^{J+M} x_2^{J-M}}{\sqrt{(J+M)!\,(J-M)!}}$$

$$= (2J)! \sum U_M^J X_M^J$$

is also invariant, hence the $U_M^J$ are also contravariant to the $X$. It follows that the $W_M^J$ are transformed just as the $U_M^J$, which is just what we wanted to prove.

The calculation of $P$ goes as follows:

$$(u_1 v_2 - u_2 v_1)^\lambda = \sum_\nu \binom{\lambda}{\nu} (-1)^\nu (u_1 v_2)^{\lambda - \nu} (u_2 v_1)^\nu$$

$$(u_1 x_1 + u_2 x_2)^{2j - \lambda} = \sum_\alpha \binom{2j - \lambda}{\alpha} (u_1 x_1)^{2j - \lambda - \alpha} (u_2 x_2)^\alpha,$$

$$(v_1 x_1 + v_2 x_2)^{2j' - \lambda} = \sum_\beta \binom{2j' - \lambda}{\beta} (v_1 x_1)^{2j' - \lambda - \beta} (v_2 x_2)^\beta.$$

Multiplying these three expressions one finds

(21.8) $$P = \lambda! (2j - \lambda)! (2j' - \lambda)! \sum_m \sum_{m'} c_{mm'}^J U_m V_{m'} X_{m+m'}^J,$$

(21.9)
$$c_{mm'}^J = \sum_\nu (-1)^\nu \frac{\sqrt{(j+m)! (j-m)! (j'+m')! (j'-m')!}}{(j-m-\nu)! (j+m-\lambda+\nu)! (j'+m'-\nu)!}$$

$$\cdot \frac{\sqrt{(J+M)! (J-M)!}}{(j'-m'-\lambda+\nu)! \nu! (\lambda-\nu)!} \quad [M = m + m'].$$

The fraction on the right in (21.9) is to be set equal to zero whenever one of the numbers $(j + m - \nu)$, etc. in the numerator becomes negative; also, $0! = 1$.

The coefficient of $X_M^J$ in (21.8) is

(21.10) $$W_M^J = \alpha_J \sum_{m+m'=M} c_{mm'}^J U_m V_{m'}$$

Since, for every fixed $J$, the $W_M^J$ can be multiplied by an arbitrary common factor, the factor $\alpha_J$ in (21.10) can be chosen arbitrarily, for example, in such a way that the $W_M^J$ form a *normed* orthogonal system in the unitary vector space generated by the products $U_m V_{m'}$. Now setting $b_{mm'}^J = \alpha_J c_{mm'}^J$, we find that the $b_{mm'}^J$ with $m + m' = M$ form for each fixed $M$ a unitary matrix $B_M$, in which $J$ occurs as column index and $m$ or $m'$ as row index. The inverse matrix $B_M^{-1}$ is, according to (7.5), simply the inverted complex conjugate matrix $B_M$. That is, the equations (21.2) can be solved for the $U_m V_{m'}$ as follows:

(21.11) $$U_m V_{m'} = \sum_J \alpha_J c_{mm'}^J W_{m+m'}^J.$$

Table (21.12)

| $J$ | $j' = \tfrac{1}{2}$ | |
|---|---|---|
|  | $m' = \tfrac{1}{2}$ | $m' = -\tfrac{1}{2}$ |
| $j+\tfrac{1}{2}$ | $\sqrt{j+m+1}$ | $\sqrt{j-m+1}$ |
| $j-\tfrac{1}{2}$ | $-\sqrt{j-m}$ | $+\sqrt{j+m}$ |

| $J$ | $j' = 1$ | | |
|---|---|---|---|
|  | $m' = 1$ | $m' = 0$ | $m' = -1$ |
| $j+1$ | $\sqrt{\dfrac{(j+m+2)(j+m+1)}{2}}$ | $\sqrt{(j+m+1)(j-m+1)}$ | $\sqrt{\dfrac{(j-m+2)(j-m+1)}{2}}$ |
| $j$ | $-\sqrt{\dfrac{2(j+m+1)(j-m)}{2}}$ | $+2m$ | $+\sqrt{\dfrac{2(j+m)(j-m+1)}{2}}$ |
| $j-1$ | $\sqrt{\dfrac{(j-m)(j-m-1)}{2}}$ | $-\sqrt{(j+m)(j-m)}$ | $\sqrt{\dfrac{(j+m)(j+m-1)}{2}}$ |

The development (21.11) is equivalent to the so-called *Clebsch-Gordan series* in the Theory of Binary Invariants[9].

In the special case $J = j + j'$ equation (21.9) simplifies to

$$c_{mm'}^J = \sqrt{\frac{(J+M)!\,(J-M)!}{(j+m)!\,(j-m)!\,(j'+m')!\,(j'-m')!}}$$

and likewise in the special case $J = j - j'$ it becomes

$$c_{mm'}^J = (-1)^{j'+m'} \sqrt{\frac{(j+m)!\,(j-m)!}{(j'+m')!\,(j'-m')!\,(J+M)!\,(J-M)!}}$$

The values of $c_{mm'}^J$ in the simplest cases are collected in the table at the left.

## C. Applications of (21.1)

Let the state of an $f$-electron system in a centrally symmetric field be given by a function $\psi(q_1, q_2, ..., q_f)$. The linear space of eigenfunctions of an energy level is transformed linearly into itself under a rotation; it thus decomposes into subspaces which are transformed according to representations $\varrho_J$. In this case one usually writes $L$ in place of $J$. Because of the uniqueness of the representation, only integer values of $L$ come into consideration.

The operator for an infinitesimal rotation of all electrons about the $z$-axis is

$$I_z = -\sum_1^f \left( x \frac{\partial}{\partial y} - y \frac{\partial}{\partial x} \right).$$

This means that $\hbar i\, I_z = \hbar L_z$ is precisely the operator for the $z$-component of the angular momentum vector $\hbar \mathscr{L}$. For an irreducible family of eigenfunctions transformed according to $\varrho_L$, the operator $\mathscr{L}^2$ has the eigenvalue $L(L+1)$, and $L_z$ has the eigenvalues $M = L, L-1, ..., -L$ (see § 20). One thus imagines in the "vector diagram" an angular momentum vector of length $\hbar L$, whose $z$-component assumes the values $\hbar M$ ($M = L, ..., -L$), exactly as in the case of the one-electron-problem (§ 8). One speaks of $S$-, $P$-, $D$-, $F$-,... terms of the atom (in analogy to the $s$-, $p$-, $d$-, $f$-,... terms of the electron) for $L = 0, 1, 2, 3, ...$ and calls $L$ the *azimuth quantum number*.

If the force potential is continuously varied (for example, if the mutual repellance of the electrons is gradually reduced) the quantum number $L$ cannot change since the representation changes continuously,

---

[9] The best textbook I know is: J. H. Grace and A. Young: The Algebra of Invariants. Cambridge: 1903.

which implies that the eigenvalue of $\mathscr{L}^2$ cannot change discontinuously. One can thus set up the possible values of $L$, initially excluding the mutual interactions between the electrons, or even better, replacing the interaction by a suitably chosen shielding of the nuclear field (see § 6) and can then slowly re-introduce this mutual interaction.

Take for example the case of two electrons. The state of each is given by a wave function $\psi_{nl}^{(m)}$ to which belongs, according to § 6, a certain term symbol $ns$ or $np$, $nd$,... according to the value of $l$. If interaction between the electrons is neglected, the eigenfunctions of the entire system are products $\psi_{nl}^{(m)}(q_1)\,\psi_{n'l'}^{(m')}(q_2)$, which are transformed according to $\varrho_l \times \varrho_{l'}$ under rotations. The reduction of this representation yields a series of subspaces, which are transformed according to $\varrho_L (L = l + l'$, $l + l' - 1,..., |l - l'|)$. If we now introduce the interaction, the atomic terms belonging to different $L$-values may separate. However, the $(2L + 1)$-tuple degeneration of the individual terms is not removed and the representations $\varrho_L$ are just the same after the re-introduction of the interaction.

In the "vector diagram" one has to combine the two vectors of length $\hbar l$ and $\hbar l'$ in such a way that the length $\hbar L$ of the resultant is either $\hbar(l + l')$ or an integer multiple of $\hbar$ less: the smallest value of the resultant is $\hbar \cdot |l - l'|$ according to (21.1).

If, for example, $l$ and $l'$ are both $= 1$ (two $p$-electrons), and if the energy levels of the electrons without interaction are $E_1$ and $E_2$, then the combination has energy $E_1 + E_2$. This term can, because of the interaction, split up into 3 terms with $L = 0, 1, 2$, hence, into an $S$-, $P$-, and $D$-term. One proceeds in the same way in all other cases.

If we start with more than two electrons, we need only apply the formula repeatedly. In the case of an $s$-, a $p$-, and a $d$-electron, for example, the calculation goes as follows:

$$\varrho_0 \times \varrho_1 \times \varrho_2 = (\varrho_0 \times \varrho_1) \times \varrho_2 = \varrho_1 \times \varrho_2 = \varrho_3 + \varrho_2 + \varrho_1$$

thus, an $F$-, $D$-, and $P$-term can arise.

The complete symbol for a term consists of the symbols of the individual electrons and that of the entire term. Consider, for example, a three-electron system and suppose that two of those electrons are in the $1\,s$-state and the third in the $2\,p$-state. The term arising is necessarily a $P$-term, so we have the symbol $1\,s^2\,2p\,P$.

An $S$-state is by definition always spherically symmetric: the $\psi$-function remains invariant under any rotation. The addition of an $s$-electron does not alter the possible values of $L$, for $\varrho_l \times \varrho_0 = \varrho_l$.

Similar considerations lead to an exact justification of the rules derived by approximate arguments in § 6 for the terms of an atom such as Li, Na, or K consisting of an outer electron and a spherically symmetric core. Previously, we replaced the mutual interaction between outer electron and core by a simple shielding of the filed of the nucleus and found $l = 0, 1, 2, \ldots$ as possible values of the angular momentum of the outer electron. If we now assume that the core is spherically symmetric in the absence of the outer electron, then for the entire system (with shielding instead of interaction) we find the value $L = l$. After introduction of the perturbation (interaction minus shielding), there occurs no splitting, but rather each term remains $(2l + 1)$-tuply degenerate and the eigenfunctions are still transformed according to $\varrho_l$. In the next section we shall see that the selection rule $l \to l \pm 1$, which justifies the classification of the terms into series, holds exactly.

### D. The Reflection Character

The field of a simple nucleus remains invariant not only under spatial rotations, but also under reflections. All reflections can be constructed from rotations and a single "reflection at the origin" or "inversion"

$$x' = -x, \quad y' = -y, \quad z' = -z,$$

which is interchangeable with all rotations. This reflection and its square (the identity) form an Abelian group of order 2. Because of the interchangeability, this Abelian group can be reduced along with the rotation group. i.e., the basis vectors of a representation (in particular, the eigenfunctions of any energy level) can always be so chosen that they are transformed according to $\varrho_L$ under a rotation and at the same time assume the factor $w = \pm 1$ under a reflection $s$. This factor $w$ is called the *reflection character*.

In particular, the spherical harmonics of order $l$ in the one-electron-problem belong to the reflection character $(-1)^l$.

If we now combine $f$ electrons in the centrally-symmetric field with azimuthal quantum numbers $l_1, l_2, \ldots, l_f$, initially neglecting the interaction, we obtain the products

$$\psi = \psi_1(q_1) \, \psi_2(q_2) \ldots \psi_f(q_f),$$

which obviously have the reflection character

$$(21.13) \qquad\qquad w = (-1)^{l_1 + l_2 + \cdots + l_f}.$$

This character remains the same after introduction of the interaction, although the eigenfunctions are then no longer products $\psi_1 \psi_2 \ldots \psi_f$. The terms thus arising are called *even* or *odd* according as $w$ is $+1$ or $-1$.

## § 22. Selection and Intensity Rules

We begin with two group-theoretic lemmas.

**Lemma 1.** *Let two representations $\varrho$, $\varrho'$ of a group $\mathscr{G}$ in the spaces $\mathscr{V} = (u_1, \ldots, u_n)$ and $\mathscr{V}' = (v_1, \ldots, v_n)$ be given by exactly the same formulae:*

$$a u_\mu = \sum_\lambda u_\lambda \alpha_{\lambda\mu}$$

$$a v_\mu = \sum_\lambda v_\lambda \alpha_{\lambda\mu}$$

*with the difference that the $u_\mu$ form a linearly independent basis for $\mathscr{V}$, while the $v_\mu$ are linearly dependent. Let the representation $\varrho$ be completely reducible*

$$(22.1) \qquad\qquad \varrho = \varrho_1 + \cdots + \varrho_k \,.$$

*Then $\varrho'$ is also completely reducible and the decomposition of $\varrho'$ is obtained by deleting some of the representations from the sum on the right in (22.1).*

*Proof.* Associating with each vector $u = \Sigma u_\lambda c_\lambda$ the vector $v = \Sigma v_\lambda c_\lambda$, we see that to the sum of two vectors $u$ corresponds the sum of the $v$, and that $av$ corresponds to $au$, so that the correspondance is an operator homomorphism. By § 13, Theorem 4, we then have

$$\mathscr{V}' \cong \mathscr{V}'_1 + \cdots + \mathscr{V}'_{h'} \,,$$

where the $\mathscr{V}'_i$ are certain irreducible subspaces occurring in the decomposition of $\mathscr{V}$. This proves the lemma.

Let us apply Lemma 1 to a product representation. We start with certain eigenfunctions $U_j^{(m)}$ and $V_j^{(m')}$ which are transformed according to $\varrho_j$ and $\varrho_{j'}$, and we want to know how the products $U_j^{(m)} \cdot V_j^{(m')}$ are transformed. If one replaces the $U$, $V$ by the same number of independent variables $u$, $v$, then the products $u_j^{(m)} \cdot v_{j'}^{(m')}$ are transformed according to $\varrho_j \times \varrho_{j'} = \sum_J \varrho_J [J = j + j', \ldots, |j - j'|]$. If one replaces the $u, v$ again by $U, V$ in the formulae of these transformations, the formulae remain valid, although the products may become linearly dependent. Our Lemma then says that they are transformed according to a representation $\Sigma \varrho_J$, in which *some* of the possible $J = j + j', \ldots, |j - j'|$ (possibly all) occur.

**Lemma 2.** *If a complete orthogonal system*

$$(22.2) \qquad\qquad \varphi_1^{(1)}, \ldots, \varphi_1^{(h)}; \quad \varphi_2^{(1)}, \ldots, \varphi_2^{(h')}; \ldots$$

*is determined in such a way that for each $\lambda$ the functions $\varphi_\lambda^{(1)}$, $\varphi_\lambda^{(2)}, \ldots$ suffer an irreducible representation $\varrho_\lambda$ under a given transformation group $\mathscr{G}$,*

*and if another set of functions $\psi^{(1)},\ldots,\psi^{(n)}$, which suffer a completely
reducible representation $\varrho$ under the same group, is expanded according
to the orthogonal system (22.2), then only those $\varphi_\lambda^{(v)}$ whose representation
$\varrho_\lambda$ is a component of the representation $\varrho$ can actually occur in the expansion.*

*Proof.* Assuming that $\psi$ is a linear combination of $\psi^{(1)}$ to $\psi^{(n)}$, let

$$(22.3) \qquad \psi = \sum_1^h a_{1v}\varphi_1^{(v)} + \sum_1^{h'} a_{2v}\varphi_2^{(v)} + \cdots = \omega_1 + \omega_2 + \cdots$$

be the expansion of $\psi$. Since $\psi$ uniquely determines all components $a_{\lambda v}$,
the $\omega_1, \omega_2$, etc. are also uniquely determined by $\psi$.

The mapping $\psi \to \omega_1$ is a linear transformation of the vector space
$\mathscr{W}$ generated by $\psi^{(1)},\ldots,\psi^{(n)}$ into the vector space $\mathscr{V}_1$ generated by
$\varphi_1^{(1)},\ldots,\varphi_1^{(h)}$, for to a sum of two vectors $\psi$ in $\mathscr{W}$ corresponds the sum of
their image vectors $\omega_1$, and to $\alpha\psi$ corresponds $\alpha\omega_1$, where $\alpha$ is any
constant. Furthermore, one can apply a transformation $t$ belonging
to the group $\mathscr{G}$ to (22.3), thus obtaining

$$t\psi = t\omega_1 + t\omega_2 + \cdots$$

where $t\omega_1$ is again an element of $\mathscr{V}_1$, etc. Hence, our mapping $\psi \to \omega_1$
maps $t\psi$ into $t\omega_1$: it is an *operator homomorphism* of $\mathscr{W}$ into $\mathscr{V}_1$. Since
$\mathscr{V}_1$ is irreducible, the images $\omega_1$ either form the whole subspace $\mathscr{V}_1$ or
they are all zero. In the former case we have an operator homomorphism
of $\mathscr{W}$ on to $\mathscr{V}_1$. By Theorem 4 of § 13, $\mathscr{V}_1$ must be isomorphic to one of
the irreducible components in the decomposition of $\mathscr{W}$, i.e. the representation $\varrho_1$ is a component of the representation $\varrho$. The same thing
holds for all representations $\varrho_\lambda$: either the $\omega_\lambda$ are zero, or the representation $\varrho_\lambda$ occurs in $\varrho$.

**Corollary** to Lemma 2. If we replace $\psi$ successively by the functions
$\psi^{(1)},\ldots,\psi^{(n)}$ and accordingly provide the $a_\lambda$ with a superscript $\mu = 1, 2,\ldots,n$,
then the coefficients $a_1^{(\mu)}$ (and likewise $a_2^{(\mu)}$, etc.) are uniquely determined
but for a factor $\lambda$ by the representation $\varrho_1$, provided no two equivalent irreducible components occur in the decomposition of the representation $\varrho$.

*Proof.* According to the proof just given, the $a_{1v}^{(\mu)}$ are the matrix elements
of a homomorphic mapping of $\mathscr{W}$ into $\mathscr{V}_1$. Now let $\mathscr{W}$ be decomposed
into its irreducible subspaces

$$\mathscr{W} = \mathscr{W}_1 \oplus \mathscr{W}_2 \oplus \cdots \oplus \mathscr{W}_k.$$

We may suppose that in $\mathscr{W}_1$ the representation $\varrho_1$ takes place, and
that the other subspaces $\mathscr{W}_2,\ldots$ are not isomorphic to $\mathscr{W}_1$. Then the
mapping $\psi \to \omega_1$ maps all $\mathscr{W}_2,\ldots$ into zero. Moreover, the matrix of

the homomorphism $\mathscr{W}_1 \to \mathscr{V}_1$ is, according to Schur's Lemma, a multiple $\lambda I$ of the identity $I$. This proves our Corollary.

The *selection rules* are based on Lemma 2. In §6 and §8 we derived the selection rules for a single electron

$$l \to l \pm 1 \quad \text{for a centrally symmetric field},$$

$$m \to m \text{ or } m \pm 1 \quad \text{for an axially symmetric field}.$$

The rule for $m$ also holds, due to its derivation, in the case of several electrons. We now investigate what happens to the rule for $l$, when $l$ is replaced by $L$.

According to §3, the intensities of the transmitted lines depend on the coefficients $a, b, c$ in the expansions

$$(22.4) \quad \begin{cases} X\psi_L^{(m)} = \sum_{L',m'} \psi_{L'}^{(m')} a_{L'L}^{(m'm)}, \\[2mm] Y\psi_L^{(m)} = \sum_{L',m'} \psi_{L'}^{(m')} b_{L'L}^{(m'm)}, \\[2mm] Z\psi_L^{(m)} = \sum_{L',m'} \psi_{L'}^{(m')} c_{L'L}^{(m'm)}. \end{cases}$$

In the notation of §21B the quantities on the left, or rather their linear combinations

$$(22.5) \qquad -(X+iY)\,\psi_L^{(m)}, \quad (X-iY)\,\psi_L^{(m)}, \quad \sqrt{2}\,Z\psi_L^{(m)}$$

are products $V_1^{(-1)}U_L^{(m)}, V_1^{(1)}U_L^{(m)}, V_1^{(0)}U_L^{(m)}$, which are transformed according to $\Sigma \varrho_{L'}$, where $L'$ can assume some of the values $L \pm 1$ and $L$. Hence, only those $\varrho_L$ can occur on the right in (22.4). This yields the selection rule

$$(22.6) \qquad L \to \begin{cases} L-1 \\ L \qquad (0 \to 0 \text{ forbidden}). \\ L+1 \end{cases}$$

In exactly the same way, but much more easily, one obtains the following selection rule for the reflection character $w = (-1)^{\Sigma l_\nu}$

$$(22.7) \qquad\qquad w \to -w$$

or *Laporte's rule: $\Sigma\, l_\nu$ jumps only by an odd number.* Indeed, if in (22.4) the $\psi_l^{(m)}$ assume the factor $w$ under a reflection $s$, then the left hand sides assume the factor $-w$ so that on the right only terms of reflection character $-w$ occur. It also follows from this rule that in the case of an outer electron with spherically symmetric core—provided the outer electron performs a quantum jump and the quantum numbers $l_\nu$ of the electrons in the core remain unchanged—, the transition $L \rightarrow L$ still admissible in (22.6) is excluded. Hence, in this case the selection rule $L \rightarrow L \pm 1$ holds, or, what is the same, $l \rightarrow l \pm 1$.

It also follows from the corollary to Lemma 2 that the coefficients $a_{L'L}^{(m'm)}$, etc. in (22.4) are uniquely determined up to a factor $\alpha_{L'L}$ independent of $m$ and $m'$ for each fixed pair of values $L, L'$. The calculation of these coefficients yields information on the intensity ratios of the lines which arise when the degeneration of the terms is removed by a noncentrally symmetric perturbation (Zeeman or Stark effect) under the assumption that the disturbance is so weak that the $\psi$-function of the unperturbed system can be used to calculate the intensity. The calculation becomes quite easy when we note that in (21.11) an expansion for the products $U_m V_{m'}$ is available. In our case we have $j = L$ and $j' = 1$, hence the factors $U_m$ and $V_{m'}$ are transformed under rotations just as our $U_L^{(m)} = \psi_L^{(m)}$ and $V_1^{(1)} = -(X + iY)$, $V_1^{(-1)} = X - iY$, $V_1^{(0)} = Z\sqrt{2}$. Hence, according to the Corollary to Lemma 2, the coefficients in the expansion of the products $U_L^{(m)} \cdot V_1^{(m')}$ must coincide with the coefficients of Table (21.12) for each $L'$ up to a common factor.

In order to bring the notations into better agreement, we replace the symbols $L, L', m'$ in (22.4) by $j, J, M$ and write

$$(22.8) \quad \begin{cases} -(X + iY)\,\psi_j^{(m)} = -\Sigma\,\psi_J^M (a + ib)_{Jj}^{(Mm)}\,, \\[2mm] (X - iY)\,\psi_j^{(m)} = \Sigma\,\psi_J^M (a + ib)_{Jj}^{(Mm)}\,, \\[2mm] 2Z\psi_j^{(m)} = \Sigma\,\psi_J^M \sqrt{2}\, c_{Jj}^{(Mm)}\,. \end{cases}$$

Now the coefficients on the right must be proportional to the expansion coefficients $c_{m,M-m}^J$ of (21.11) for each $J$. This yields (see the table in § 21 B):

$$(22.9) \quad \begin{cases} \text{for } J = j+1: & (a + ib)_{J,j}^{(m+1,m)} = -\alpha\sqrt{(j + m + 2)(j + m + 1)}\,, \\[2mm] & (a - ib)_{J,j}^{(m-1,m)} = \alpha\sqrt{(j - m + 2)(j - m + 1)}\,, \\[2mm] & c_{J,j}^{(m,m)} = \alpha\sqrt{(j + m + 1)(j - m + 1)}\,; \\[3mm] \text{for } J = j: & (a + ib)_{J,j}^{(m+1,m)} = \beta\sqrt{(j + m + 1)(j - m)}\,, \\[2mm] & (a - ib)_{J,j}^{(m-1,m)} = \beta\sqrt{(j + m)(j - m + 1)}\,, \\[2mm] & c_{J,j}^{(m,m)} = \beta m\,; \\[3mm] \text{for } J = j-1: & (a + ib)_{J,j}^{(m+1,m)} = \gamma\sqrt{(j - m)(j - m - 1)}\,, \\[2mm] & (a - ib)_{J,j}^{(m-1,m)} = -\gamma\sqrt{(j + m)(j + m - 1)}\,, \\[2mm] & c_{J,j}^{m,m} = \gamma\sqrt{(j + m)(j - m)}\,. \end{cases}$$

Subsequently, the $j$, $J$ are to be replaced by $L$, $L'$ in these formulae. If desired, one can also calculate $a_{Jj}^{Mm}$ and $b_{Jj}^{Mm}$ from $(a+ib)_{Jj}^{Mm}$ and $(a-ib)_{Jj}^{Mm}$. The squares of these numbers yield by §3 the transition probabilities to which the intensities of the corresponding spectral lines are proportional. The direction of polarisation of the emitted light has already been given in § 8.

## § 23. The Representations of the Lorentz-Group

### A. The Group SL(2) and the Restricted Lorentz Group

In the same way as the representations of the rotation group $\mathcal{O}_3$ were derived in § 19 and § 20, we now want to obtain the representations of the group of Lorentz transformations (Lorentz group).

We start with the group SL(2) of linear transformations of determinant 1 in a two-dimensional complex vector space. Since we will deal later with co- and contra-variant vectors, we make use of the Ricci-calculus, denoting the basis vectors by $\overset{1}{u}, \overset{2}{u}$ and their linear combinations by $a_1 \overset{1}{u} + a_2 \overset{2}{u}$. The transformation formulae are then

$$(23.1) \qquad \begin{cases} \overset{1}{u}' = \overset{1}{u}\alpha + \overset{2}{u}\gamma, \\ \overset{2}{u}' = \overset{1}{u}\beta + \overset{2}{u}\delta, \end{cases} \quad \alpha\delta - \beta\gamma = 1.$$

We consider a second vector space $(a_{1\cdot}\overset{\dot{1}}{u} + a_{2\cdot}\overset{\dot{2}}{u})$, which will be transformed along with the first but each time with the matrix of complex conjugates:

$$(23.2) \qquad \begin{cases} \overset{\dot{1}}{u}' = \overset{\dot{1}}{u}\alpha^* + \overset{\dot{2}}{u}\gamma^*, \\ \overset{\dot{2}}{u}' = \overset{\dot{1}}{u}\beta^* + \overset{\dot{2}}{u}\delta^*, \end{cases}$$

We agree to write all quantities transformed according to (23.2) with dotted indices $\dot{1}, \dot{2}$.

If $(a_1, a_2)$ and $(b_1, b_2)$ are two vectors both transformed according to (23.1), then $a_1 b_2 - a_2 b_1$ is invariant; consequently, the vector $(b_2, -b_1)$ is contravariant to $(a_1, a_2)$. Hence, we can form from each binary vector $(b_1, b_2)$ a contravariant vector $(b^1, b^2)$ with components

$$(23.3) \qquad\qquad b^1 = b_2, \qquad b^2 = -b_1.$$

In the same way we define for every vector $(b_{1\cdot}, b_{2\cdot})$ a contravariant vector $(b^{\dot{1}}, b^{\dot{2}}) = (b_{2\cdot}, -b_{1\cdot})$.

The linear space of all bilinear forms

$$(23.4) \qquad\qquad c_{1\dot{1}} \cdot \overset{1}{u}\overset{\dot{1}}{u} + c_{1\dot{2}} \cdot \overset{1}{u}\overset{\dot{2}}{u} + c_{2\dot{1}} \cdot \overset{2}{u}\overset{\dot{1}}{u} + c_{2\dot{2}} \cdot \overset{2}{u}\overset{\dot{2}}{u}$$

is linearly transformed into itself by the transformations (23.1) and (23.2). The determinant

$$c_{11} \cdot c_{22} \cdot - c_{12} \cdot c_{21} \cdot$$

remains invariant, as one can verify by direct calculation.

We now assume that (23.4) is a Hermitean form in the sense of § 9, i.e. that $c_{11}$ and $c_{22}$ are real, while $c_{12}$ and $c_{21}$ are complex conjugates. This assumption is invariant under the transformations (23.1) and (23.2). The Hermitean forms form a real four-dimensional vector space. In this space we can introduce real coordinates $x, y, z, t$ by putting

(23.5)
$$\begin{cases} c_{21} \cdot = x + iy \\ c_{12} \cdot = x - iy \\ c_{11} \cdot = z + ct \\ c_{22} \cdot = -z + ct \end{cases}$$

Under the group SL(2), the real coordinates $x, y, z, t$ undergo real linear transformations, and the quadratic form

(23.6)       $$c_{11} \cdot c_{22} \cdot - c_{12} \cdot c_{21} \cdot = c^2 t^2 - z^2 - x^2 - y^2$$

remains invariant. Hence our transformations are real Lorentz transformations of the variables $x, y, z, t$.

Putting the quadratic form (23.6) equal to zero, one obtains the *light cone*

$$c^2 t^2 - x^2 - y^2 - z^2 = 0$$

This cone divides the 4-space into three parts. Inside the light cone, on the positive and negative side ($t > 0$ and $t < 0$ respectively) the quadratic form is positive, outside it is negative. The inner points on the positive side represent positive Hermitean forms. The group SL(2) transforms positive forms into positive ones, hence the transformations of the variables $x, y, z, t$ do not change the direction of time. A simple calculation shows that the determinants of these transformations are $+1$. Hence the transformations of the coordinates $x, y, z, t$ induced by the group SL(2) do not change the orientation of the $(x, y, z)$-space: they belong to the *restricted Lorentz group*.

The points on the positive light cone represent semi-definite Hermitean forms, which can be written as products

$$(s_1 \overset{1}{u} + s_2 \overset{2}{u})(s_1^* \overset{\dot 1}{u} + s_2^* \overset{\dot 2}{u}).$$

Their coefficients are

(23.7)
$$\begin{cases}
x + iy = c_{21'} = s_2 s_1^* \\
x - iy = c_{12'} = s_1 s_2^* \\
z + ct = c_{11'} = s_1 s_1^* \\
-z + ct = c_{22'} = s_2 s_2^* .
\end{cases}$$

Thus we obtain a parametric representation of the positive light cone. We might have started with this parametric representation and derive everything else from it.

We now prove that every transformation of the restricted Lorentz group can be obtained from a transformation of SL(2).

It is easy to show that all spatial rotations can be obtained. Indeed, if one chooses $\delta = \alpha^*$ and $\gamma = -\beta^*$ in (23.1) and (23.2), the sum

$$2\,ct = c_{11'} + c_{22'}.$$

remains invariant, the transformations (23.1) become unitary and the variables $x, y, z$ undergo exactly the transformations given in § 19.

On the other hand, among our Lorentz transformations the following also occur

$$\overset{1}{u}' = \alpha\, \overset{1}{u} \qquad\qquad (\alpha > 1)$$

$$\overset{2}{u}' = \alpha^{-1}\, \overset{2}{u}$$

(23.8)
$$\begin{cases}
c_{11'}' = \alpha^2 c_{11'} \\
c_{22'}' = \alpha^{-2} c_{22'} \\
c_{12'}' = c_{12'} \\
c_{21'}' = c_{21'}
\end{cases}
\qquad
\begin{cases}
x' = x \\
y' = y \\
z' = \tfrac{1}{2}(\alpha^2 + \alpha^{-2})\,z + \tfrac{1}{2}(\alpha^2 - \alpha^{-2})\,ct \\
ct' = \tfrac{1}{2}(\alpha^2 - \alpha^{-2})\,z + \tfrac{1}{2}(\alpha^2 + \alpha^{-2})\,ct .
\end{cases}$$

The world-line of a point at rest is transformed by this transformation into the world-line of a point moving with arbitrary velocity

$$v = c\,\frac{\alpha^4 - 1}{\alpha^4 + 1}$$

in the $z$-direction. It is known from Special Relativity Theory that these special Lorentz transformations together with all rotations generate the restricted Lorentz group. Hence:

*The restricted Lorentz group can be obtained as a representation of the group* SL(2).

The only transformations (23.1) which yield the identity of the Lorentz group are given by the matrices

$$I = \begin{pmatrix} 1 & 0 \\ 0 & 1 \end{pmatrix} \quad \text{and} \quad -I = \begin{pmatrix} -1 & 0 \\ 0 & -1 \end{pmatrix}$$

Hence we obtain the final result:

*The group* SL(2) *is a two-valued representation of the restricted Lorentz group.*

## B. Infinitesimal Transformations

We now want to determine all differentiable representations of the restricted Lorentz group by using the method of infinitesimal transformations.

Every representation of the restricted Lorentz Group is also a representation of the group SL(2). Conversely, every representation of SL(2) is a (at most two-valued) representation of the restricted Lorentz group. Hence, we first seek representations of SL(2).

The matrix of a transformation in SL(2) can be written as

$$(23.9) \qquad A = \begin{pmatrix} \alpha & \beta \\ \gamma & \delta \end{pmatrix} = \begin{pmatrix} 1 + \alpha_1 + i\alpha_2 & \alpha_3 + i\alpha_4 \\ \alpha_5 + i\alpha_6 & \delta \end{pmatrix}$$

with

$$\delta = \frac{1 + \beta\gamma}{\alpha} = \frac{1 + (\alpha_3 + i\alpha_4)(\alpha_5 + i\alpha_6)}{1 + \alpha_1 + i\alpha_2}.$$

As parameters in the neighborhood of unity we may use the real variables $\alpha_1, \ldots, \alpha_6$. As in § 20 we define the infinitesimal transformations $I_1, \ldots, I_6$ of an arbitrary representation and find, exactly as before, that there must exist commutation relations of the form

$$I_\mu I_\nu - I_\nu I_\mu = \sum_\sigma I_\sigma c_{\lambda\mu}^\sigma.$$

The $c_{\lambda\mu}^\sigma$ are real numbers depending only on the composition of the group and can thus be determined from any faithful representation, for example, from the matrices of SL(2) itself. For this representation we have from (23.9)

$$I_1 = \frac{\partial A}{\partial \alpha_1} = \begin{pmatrix} 1 & 0 \\ 0 & -1 \end{pmatrix}; \quad I_3 = \begin{pmatrix} 0 & 1 \\ 0 & 0 \end{pmatrix}; \quad I_5 = \begin{pmatrix} 0 & 0 \\ 1 & 0 \end{pmatrix}$$

$$I_2 = \frac{\partial A}{\partial \alpha_2} = \begin{pmatrix} i & 0 \\ 0 & -i \end{pmatrix}; \quad I_4 = \begin{pmatrix} 0 & i \\ 0 & 0 \end{pmatrix}; \quad I_6 = \begin{pmatrix} 0 & 0 \\ i & 0 \end{pmatrix}.$$

One finds the following commutation relations with real coefficients[10]:

$$I_1 I_3 - I_3 I_1 = \quad 2I_3 \qquad I_1 I_4 - I_4 I_1 = \quad 2I_4 \qquad I_2 I_3 - I_3 I_2 = \quad 2I_4$$

$$I_1 I_5 - I_5 I_1 = -2I_5 \qquad I_1 I_6 - I_6 I_1 = -2I_6 \qquad I_2 I_5 - I_5 I_2 = -2I_6$$

$$I_3 I_5 - I_5 I_3 = \quad I_1 \qquad I_3 I_6 - I_6 I_3 = \quad I_2 \qquad I_4 I_5 - I_5 I_4 = \quad I_2$$

$$I_2 I_4 - I_4 I_2 = -2I_3 \qquad I_1 I_2 - I_2 I_1 = 0$$

$$I_2 I_6 - I_6 I_2 = \quad 2I_5 \qquad I_3 I_4 - I_4 I_3 = 0$$

$$I_4 I_6 - I_6 I_4 = - \quad I_1 \qquad I_5 I_6 - I_6 I_5 = 0.$$

These relations must also hold for an arbitrary representation. They can be simplified by introducing the new operators

$$I_1 + iI_2 = 4A_z; \qquad I_3 + iI_4 = 2A_p; \qquad I_5 + iI_6 = 2A_q;$$

$$I_1 - iI_2 = 4B_z; \qquad I_3 - iI_4 = 2B_p; \qquad I_5 - iI_6 = 2B_q.$$

The recalculation shows that each $A$ is interchangeable with each $B$:

$$A_h B_k - B_k A_h = 0 \quad \text{for} \quad h, k = z, p, q$$

and also

$$A_z A_p - A_p A_z = \quad \cdot A_p \qquad B_z B_p - B_p B_z = \quad B_p$$

$$A_z A_q - A_q A_z = - \quad A_q \qquad B_z B_q - B_q B_z = - \quad B_q$$

$$A_p A_q - A_q A_p = \quad 2A_z \qquad B_p B_q - B_q B_p = \quad 2B_z.$$

These are the same commutation rules for $A$ as well as $B$ as in (20.2). Consequently, all of the conclusions which were obtained from (20.2) also hold. If $v_J$ is a vector corresponding to the largest eigenvalue $J$ of $A_z$, then one obtains a complete set of eigenvectors $v_M(-J \leq M \leq J)$ which are transformed by the operators $A_k$ according to (20.8) (with $A_k$ instead of $L_k$). The totality of all $v_J$ belonging to the eigenvalue $J$ is a linear space, which is invariant under the transformations $B_k$, since the latter are interchangeable with the $A_z$. By the same principle, one can find a set of vectors $v_{JM'}(-J' \leq M' \leq J')$ in this space, which are transformed under $B_k$ according to (20.5). Each of these $v_{JM'}$ yields by repeated ap-

---

[10] The fact that the coefficients must be real follows from the general considerations of § 18.

plication of the operator $A_q$ a complete set of vectors $v_{MM'} (-J \leqq M \leqq J)$. *In this way one finds* $(2J+1)(2J'+1)$ *vectors* $v_{MM'}$, *for which*

(23.10)
$$\begin{cases} A_p v_{MM'} = \sqrt{(J-M)(J+M+1)}\, v_{M+1,M'} \\ A_q v_{MM'} = \sqrt{(J+M)(J-M+1)}\, v_{M-1,M'} \\ A_z v_{MM'} = M v_{MM'} \\ B_p v_{MM'} = \sqrt{(J'-M')(J'+M'+1)}\, v_{M,M'+1} \\ B_q v_{MM'} = \sqrt{(J'+M')(J'-M'+1)}\, v_{M,M-1} \\ B_z v_{MM'} = M' v_{MM'} \end{cases}$$

*and which determine an irreducible representation of the group* SL(2). The irreducibility results easily from the same arguments which were used for this purpose in § 20. If the original representation was irreducible, then the $v_{MM'}$ must necessarily span the entire space: *this implies that each irreducible representation is equivalent to one of the representations* $D_{JJ'}$ *given by* (23.10).

It is easy to give a system of variables which is transformed precisely according to $D_{JJ'}$; we need only set

$$v_{MM'} = \frac{\overset{1}{u}{}^{J+M}\,\overset{2}{u}{}^{J+M}}{\sqrt{(J+M)!\,(J-M)!}} \cdot \frac{\overset{1'}{u}{}^{J'+M'}\,\overset{2'}{u}{}^{J'+M'}}{\sqrt{(J'+M')!\,(J'-M')!}}\,.$$

An arbitrary linear combination of these $v_{MM'}$ is given by the expression

$$\Sigma\, c_{\lambda\mu\cdots\nu\varrho\cdot\sigma\cdot\cdots\tau}\cdot \overset{\lambda}{u}\overset{\mu}{u}\ldots\overset{\nu}{u}\overset{\check{\varrho}}{u}\overset{\check{\sigma}}{u}\ldots\overset{\tau}{u}$$

where the tensor $c$ is symmetric in the $2J$ indices $\lambda, \ldots, \nu$ and symmetric in the $2J'$ indices $\varrho\cdot, \ldots, \tau\cdot$. *These tensors form the vector space of the irreducible representation* $\varrho_{JJ'}$ *of the group* SL(2).

H. Weyl was the first to prove that every representation is *completely reducible*.

The considerations above determine all "quantities" which are in some way linearly transformed under the restricted Lorentz group. The simplest of these quantities are the invariants or scalars, then come the binary vectors $(a_1, a_2)$ and $(a_{1\cdot}, a_{2\cdot})$. The contravariant vectors $(a^1, a^2)$ are transformed equivalently to the covariant ones according to (23.3). Next we have the tensors $c_{\lambda\mu\cdot}$, which are equivalent to the world vectors by (23.5), next the symmetric tensors $c_{\lambda\mu}$ and $c_{\lambda\cdot\mu\cdot}$ with three components, etc.

The collective designation "*spinors*" has been given to these quantities because of the role they play in the theory of the spinning electron.

The notation of spinors by means of dotted and undotted indices was first introduced in my paper "Spinor Analysis", Nachrichten Ges. der Wiss. Göttingen 1929, p. 100. By means of this notation, all invariant relations between world vectors and spinors can be written in such a form that the invariance becomes evident. For examples see the next Section $C$ and also § 27.

## C. The Relation between World Vectors and Spinors

From now on, the four components of a world-vector in Minkowski space will be denoted as follows:

$$x^0 = ct, \quad x^1 = x, \quad x^2 = y, \quad x^3 = z.$$

The covariant components of the same vector are, in Einstein's notation,

$$x_0 = -ct, \quad x_1 = x, \quad x_2 = y, \quad x_3 = z.$$

The formulae (23.5) now read

(23.11)
$$\begin{cases} c_{21^{\cdot}} = x^1 + ix^2 \\ c_{12^{\cdot}} = x^1 - ix^2 \\ c_{11^{\cdot}} = x^3 + x^0 \\ c_{22^{\cdot}} = -x^3 + x^0. \end{cases}$$

These formulae may also be written in matrix form

(23.12)
$$C = \begin{pmatrix} c_{11^{\cdot}} & c_{12^{\cdot}} \\ c_{21^{\cdot}} & c_{22^{\cdot}} \end{pmatrix} = \begin{pmatrix} x^3 + x^0 & x^1 - ix^2 \\ x^1 + ix^2 & -x^3 + x^0 \end{pmatrix}$$

$$= x^0 \cdot \begin{pmatrix} 1 & 0 \\ 0 & 1 \end{pmatrix} + x^1 \cdot \begin{pmatrix} 0 & 1 \\ 1 & 0 \end{pmatrix} + x^2 \cdot \begin{pmatrix} 0 & -i \\ i & 0 \end{pmatrix} + x^3 \cdot \begin{pmatrix} 1 & 0 \\ 0 & -1 \end{pmatrix},$$

$$C = x^0 s_0 + x^1 s_1 + x^2 s_2 + x^3 s_3.$$

If the elements of the four matrices $s_k$ are denoted by $\sigma_{k\lambda\nu^{\cdot}}$, we may also write (23.12) as

(23.13)
$$c_{\lambda\nu^{\cdot}} = \Sigma \, x^k \sigma_{k\lambda\nu^{\cdot}}.$$

Thus, to every world-vector $x^k$ corresponds a spinor $c_{\lambda\mu^{\cdot}}$ and conversely. If the world-tensor is real, the spinor defines a Hermitean form and the matrix $C$ is self-adjoint, i.e. $c_{11'}$ and $c_{22'}$ are real and $c_{12'}$ complex conjugate to $c_{21'}$, and conversely. The relation (23.13) is invariant under Lorentz-Transformations. In § 23 A we have seen that to every covariant vector $(b_1, b_2)$ a contravariant vector $(b^1, b^2)$ can be defined according to the formulae (23.3):

$$b^1 = b_2, \quad b^2 = -b_1.$$

These formulae may also be written as

(23.14)
$$b^\lambda = \Sigma \, \varepsilon^{\lambda\mu} b_\mu$$

with

$$\varepsilon^{12} = 1, \quad \varepsilon^{21} = -1, \quad \varepsilon^{11} = \varepsilon^{22} = 0.$$

Just so, to every dotted vector $(b_{1'}, b_{2'})$ a contravariant vector $(b^{1'}, b^{2'})$ may be defined by

(23.15)
$$b^{\lambda'} = \Sigma \, \varepsilon^{\lambda'\mu'} b_{\mu'}.$$

with

$$\varepsilon^{1'2'} = 1, \quad \varepsilon^{2'1'} = -1, \quad \varepsilon^{1'1'} = \varepsilon^{2'2'} = 0.$$

The tensor components $c_{\lambda\nu'}$ transform just as products $a_\lambda b_{\nu'}$, hence we may define contravariant components $c^{\kappa\mu'}$ by

(23.16)
$$c^{\kappa\mu'} = \Sigma \, \varepsilon^{\kappa\lambda} \varepsilon^{\mu'\nu'} c_{\lambda\nu'}.$$

or explicitly

(27.17)
$$c^{11'} = c_{22'}, \quad c^{12'} = -c_{21'}, \quad c^{21'} = -c_{12'},$$
$$c^{22'} = c_{11'}.$$

The matrix formed by these $c^{\kappa\mu'}$ is

$$C' = \begin{pmatrix} c^{11'} & c^{12'} \\ c^{21'} & c^{22'} \end{pmatrix} = \begin{pmatrix} c^{22'} & -c^{21'} \\ -c^{12'} & c^{22'} \end{pmatrix}$$

$$= \begin{pmatrix} -x^3 + x^0 & -x^1 - ix^2 \\ -x^1 - ix^2 & x^3 + x^0 \end{pmatrix}$$

$$= x^0 s_0 - x^1 s_1 - x^2 s_2 - x^3 s_3.$$

If the elements of the four matrices

$$s_0' = s_0, \qquad s_1' = -s_1, \qquad s_2' = -s_2, \qquad s_3' = -s_3$$

are called $\sigma_k^{\varkappa\mu\cdot}$ $(k = 0, 1, 2, 3)$, we have

(23.18) $$c^{\varkappa\mu\cdot} = \Sigma \, x^h \sigma_h'^{\varkappa\mu\cdot}.$$

This relation too is invariant under Lorentz transformations.

The formulae (27.13) and (23.18) show that every world vector $(x^k)$ determines a covariant spintensor $(c_{\lambda\nu\cdot})$ and a contravariant spintensor $(c^{\varkappa\mu\cdot})$. The relation between those two spintensors is given by (27.17). All this is just elementary algebra, but we shall need it later.

Chapter IV

# The Spinning Electron

## § 24. The Spin

We have seen in § 8 that Schrödingers Wave Equation, which includes a magnetic perturbation term $\kappa \mathscr{H} \cdot \mathscr{L}$ in the energy operator ($\kappa =$ Bohr's magneton) can only explain the "normal" Zeeman effect as it occurs in the singlet terms. In order to explain the anomalous Zeeman effect and the multiplet splitting, it thus appears indispensible to assume, in addition to the magnetic moment of the orbital motion, another magnetic moment not depending on the orbital motion. According to the hypothesis of Uhlenbeck and Goudsmit, this moment arises from the so-called *spin*, i.e. from the angular momentum of the "spinning" electron[1].

If the direction of magnetization of a ferro-magnetic bar is reversed, the bar acquires an angular momentum. The observed ratio between the changes in mechanical angular momentum and in magnetic moment is as $\hbar$ to $2\kappa$ instead of $\hbar$ to $\kappa$, as it would be if the magnetization were due to the orbital motion of the electrons. If one wants to explain this anomaly by making the electron spin responsible for ferro-magnetism, one must assume that the magnetic moment of the spinning electron is twice as large as the magnetic moment of an orbital motion having the same mechanical angular momentum.

The Stern-Gerlach experiment shows that the angular momentum of the spinning electron is *quantized*, i.e., that its component in an arbitrary direction can take on only discrete values. In this experiment, a ray of silver atoms in the ground state ($l = 0$) is emitted in the $x$-direction in a magnetic field which changes its magnitude in the $z$-direction. Such a field exerts a force

$$\frac{\partial H_z}{\partial z} M_z$$

on a magnet whose moment in the $z$-direction has the value $M_z$. The ray is split into two component rays corresponding to the values $M_z = \pm \kappa$.

---

[1] For the history of this concept see my article: Spin and Statistics. In: Pauli Memorial Volume (edited by Fierz and Weisskopf) New York: Interscience Publishers 1960.

If we make the plausible assumption that only one electron in the atom is responsible for the magnetic moment, while the spins of the other electrons cancel each other (this assumption is plausible because the silver ion $Ag^+$ shows no Zeeman effect in the ground state), we may conclude that the magnetic moment of the electron in any given direction can only take the values $\pm\kappa$. Since the ratio of the magnetic moment to the mechanical angular momentum was assumed to be as $\kappa$ is to $\frac{1}{2}\hbar$, it follows that the mechanical spin momentum in any direction is always $\pm\frac{1}{2}\hbar$.

This quantization of the spin makes it possible to explain the multiplet splitting of the spectral terms. In the simplest case of the alkali metals, where only one electron plays an essential role, the phenomenon is as follows: in first approximation the terms agree with the energy values calculated in §6 for an electron in a central field, but they all form doublets, except for the $s$-terms ($l = 0$), which remain single. If a magnetic field is introduced, one of the terms of the doublet splits into $2l + 2$ and the other into $2l$ terms, whereas in the spin-free theory a split into $2l + 1$ terms would be expected. One can distinguish the terms of the doublet by a quantum number $j$ which assumes the value $l + \frac{1}{2}$ for the term of multiplicity $2l + 2$ and the value $l - \frac{1}{2}$ for the other term, so that both terms have multiplicity $2j + 1$.

In order to make the situation clearer in a heuristic way, one imagines that the vector of orbital angular momentum, of length $\hbar l$, and the spin momentum vector of length $\frac{1}{2}\hbar$, combine to form a resultant vector of length $\hbar j$ with $j = l \pm \frac{1}{2}$. This total angular momentum $\hbar j$ determines the multiplicity $2j + 1$, just as the orbital angular momentum $\hbar l$ determines the multiplicity $2l + 1$ in the "spin-free" case. Further one imagines that the two terms $j = l + \frac{1}{2}$ and $j = l - \frac{1}{2}$ separate due to the interaction of the spin $\frac{1}{2}\hbar$ with the orbital angular momentum $\hbar l$. A justification of this "vector schema" will be given later.

Landé has found empirically that any one of the terms $j = l \pm \frac{1}{2}$ splits up in a weak magnetic field into $2j + 1$ equi-distant components, the deviations from the undisturbed terms being given by

$$(24.1) \qquad g\kappa H_z m \left( g = \frac{j + \frac{1}{2}}{l + \frac{1}{2}}; \ m = j, j - 1, \dots, -j \right).$$

In the case $l = 0$, where the entire angular momentum is due to the spin, we have $m = \pm\frac{1}{2}$ and $g = 2$. The values $m = \pm\frac{1}{2}$ are, after multiplication by $\hbar$, exactly the possible values of the $z$-component of the angular momentum, and the factor $g = 2$ again shows that to the angular momentum $\hbar m$ a magnetic moment $2\kappa m$ corresponds.

The hypotheses thus obtained are:

1. The electron has its own mechanical angular momentum or spin whose component in any fixed direction can take on only the values $\pm\frac{1}{2}\hbar$.

2. The energy effect of the spin is small compared with that of the charge and mass, provided no external magnetic field is present.

3. A magnetic moment $\kappa$ corresponds to the mechanical momentum $\frac{1}{2}\hbar$.

## § 25. The Wave Function of the Spinning Electron

### A. Pauli's Pair of Functions $(\psi_1, \psi_2)$

Let us try to reformulate these hypotheses in the language of wave mechanics. The existence of spin means that an electron is not merely a mass point having but three coordinates $x$, $y$, $z$. At least one degree of freedom due to the spin must be added. We choose for this the $z$-component of the spin angular momentum, measured in multiples of $\frac{1}{2}\hbar$. This $z$-component is a variable $\sigma_z$ which can only assume the values $+1$ and $-1$ by Hypothesis 1 of § 24. Following Pauli[2], we assume a wave function

$$\psi(x, y, z, \sigma_z) = \psi(q, \sigma_z)$$

in which the space coordinates $q$ run through the entire space, while $\sigma_z$ assumes only the values $+1$ and $-1$. This function is equivalent to the function pair

$$\psi_1 = \psi(q, 1); \qquad \psi_2 = \psi(q, -1)$$

or, as we prefer to say, it is a wave function with two "components" $\psi_1$, $\psi_2$, which are functions of the space coordinates $q$ only.

In the statistical interpretation of wave mechanics $\int \psi_1^* \psi_1 \, dq$ integrated over a volume in space, is proportional to the probability that the electron with spin directed parallel to the positive $z$-axis is encountered in this volume $V$. Likewise, $\int \psi_2^* \psi_2 \, dq$ is proportional to the probability that the electron is encountered in $V$ with reversed spin. The sum

$$\int (\psi_1^* \psi_1 + \psi_2^* \psi_2) \, dV$$

gives the total probability for the electron to be located in the volume $V$.

---

[2] W. Pauli: Zur Quantenmechanik des magnetischen Elektrons. Zeitschr. für Physik 43, p. 601 (1927).

## B. Transformation of the Pair $(\psi_1, \psi_2)$

We now ask: How are the components $\psi_1$, $\psi_2$ transformed if the state of the electron is subjected to an arbitrary space rotation $R$, or, what amounts to the same thing, if the state is left unchanged and the co-ordinate system subjected to the rotation $R^{-1}$?

Pauli was the first to propose a unitary transformation

$$(25.1) \qquad \begin{cases} \psi_1' = t_{11}\psi_1 + t_{12}\psi_2 \\ \psi_2' = t_{21}\psi_1 + t_{22}\psi_2, \end{cases}$$

the matrix elements $t_{11}$, $t_{12}$, $t_{21}$, $t_{22}$ being the complex conjugates $\alpha^*, \beta^*, \gamma^*, \delta^*$ of the so-called Cayley-Klein parameters of the rotation $R^{-1}$, as defined in the book of A. Sommerfeld and F. Klein: Theorie des Kreisels, Sections 2—4. The matrix $T$ formed by these coefficients $t_{\alpha\beta}$ is just the matrix representing the rotation $R$ in the representation $\varrho_{\frac{1}{2}}$ of the rotation group $\mathcal{O}_3$.

J. von Neumann and E. Wigner proved in 1927 that the formulae (25.1) are a necessary consequence of the representation theory of the orthogonal group. In the Pauli Memorial Volume quoted at the beginning of this chapter, I have shown that one can assume less and prove more than von Neumann and Wigner did. Here I shall follow the same line of thought as in the Pauli Memorial Volume.

Starting with a definite orthogonal coordinate system, let us consider a single electron and suppose, as Pauli did, a state of the electron to be defined by a pair of functions $(\psi_1, \psi_2)$ of the space coordinates. If the state of the electron is subjected to an arbitrary space rotation $R$ the new state is represented by a pair of functions $(\psi_1', \psi_2')$. The probability for an electron to be found in an infinitesimal volume at a point $P$ is

$$\{\psi_1^* \psi_1(P) + \psi_2^* \psi_2(P)\}\, dV.$$

If $P$ is transformed into $P'$ by the rotation $R$, the probability of finding the transformed electron in $dV'$ must be equal to the probability of finding the original electron in $dV$, and since the volume $dV'$ is equal to $dV$, we must have

$$(25.2) \qquad \psi_1'^* \psi_1'(P') + \psi_2'^* \psi_2'(P') = \psi_1^* \psi_1(P) + \psi_2^* \psi_2(P).$$

It is reasonable to assume, as Pauli did, that $\psi_1'(P')$ and $\psi_2'(P')$ are linear functions of $\psi_1(P)$ and $\psi_2(P)$ with coefficients depending only on the rotation $R$:

$$(25.3) \qquad \begin{cases} \psi_1'(P') = t_{11}\psi_1(P) + t_{12}\psi_2(P) \\ \psi_2'(P') = t_{21}\psi_1(P) + t_{22}\psi_2(P). \end{cases}$$

Using one-column, two-row matrices $\Psi'$ and $\Psi$, we may write the Eqs. (25.3) in matrix form

(25.4) $$\Psi'(P') = T \, \Psi(P)$$

where $T = T(R)$ depends only on the rotation $R$. Because of (25.2), the matrix $T$ must be unitary.

If $\beta = e^{i\alpha}$ is a complex number of absolute value 1, the functions $\Psi$ and $\beta\Psi$ determine the same state, hence the matrix may be multiplied by a constant factor $\beta = e^{i\alpha}$. The set of all matrices $\beta T$, where $\beta$ varies on the unit circle in the complex plane, may be called a *projective unitary transformation* $T_{\text{proj}}$. These transformations form the *projective unitary group*.

If two space rotations $R$ and $S$ are applied one after another (first $S$, next $R$), a new rotation $RS$ is obtained. Obviously,

(25.5) $$T(RS) = \beta T(R) \, T(S) \quad (\beta = e^{i\alpha}).$$

The relation (25.5) is expressed in words by saying that the projective unitary transformations $T_{\text{proj}}$ form a "projective unitary representation" of the group $\mathcal{O}_3$.

The group $\mathcal{O}_3$ is a simple group: it has no normal divisions except itself and the unit subgroup consisting of 1 only. Now in Math. Zeitschrift **36**, p. 780, I have proved that every unitary representation of a simple Lie group is continuous. The proof holds just as well for projective unitary representations. Hence $T_{\text{proj}}$ is a continuous function of $R$.

Multiplying the matrices $T$ by suitable factors $\beta = e^{i\alpha}$, we may assume that their determinants are 1. Now the only freedom left is, to replace $T$ by $-T$. Hence (25.5) may be simplified to

(25.6) $$T(RS) = \pm T(R) \, T(S)$$

which means: The matrices $T$ form a two-valued representation of the rotation group $\mathcal{O}_3$.

For $R = 1$ we may choose $T(R) = 1$. The continuity of $T_{\text{proj}}$ now implies: if $R$ varies in a certain neighbourhood $U$ of 1 in $\mathcal{O}_3$, the matrices $T(R)$ may be multiplied by such factors $\beta = \pm 1$ that in the resulting matrices $\beta T$ all elements differ from those of the unit matrix by less than $\varepsilon$. Hence the matrix $\beta T$, which may again be called $T$, is a uniquely determined continuous function of $R$ in the neighbourhood $U$ of 1. For $R = S = 1$, the factor $\pm 1$ in (25.6) is $+ 1$, hence, by continuity, it is $+ 1$ for all $R$ and $S$ in the neighbourhood $U$.

Von Neumann has proved that such a continuous representation of a neighbourhood of 1 in a Lie group is always analytic and is determined by the matrices representing the infinitesimal transformations of the group[3]. Hence, the required two-valued representation $R \to T(R)$ is uniquely determined by the matrices $I_x, I_y, I_z$ representing the infinitesimal rotations of $\mathcal{O}_3$.

In Chapter III we have determined all possible sets of matrices $I_x, I_y, I_z$ by means of the commutation relations $I_x I_y - I_y I_x = I_z$, etc. The irreducible representations were found to be $\varrho_0, \varrho_{\frac{1}{2}}, \varrho_1, \ldots$. All unitary representations are completely reducible.

In our case we need a representation of degree 2. Hence we have to take either the trivial representation $T(R) = 1$ of degree 2 or the irreducible representation $\varrho_{\frac{1}{2}}$. Now a rotation inverting the $z$-axis must transform an eigenfunction $(\psi_1, 0)$ into an eigenfunction $(0, \psi_2)$. Hence the trivial representation will not do, and the representation $\varrho_{\frac{1}{2}}$ is the only possibility.

The infinitesimal transformations of the representation are, according to the theory of § 20,

(25.7)
$$\begin{cases} I_x = \dfrac{1}{2i} s_1 = \dfrac{1}{2i} \begin{pmatrix} 0 & 1 \\ 1 & 0 \end{pmatrix} \\[2ex] I_y = \dfrac{1}{2i} s_2 = \dfrac{1}{2i} \begin{pmatrix} 0 & -i \\ i & 0 \end{pmatrix} \\[2ex] I_z = \dfrac{1}{2i} s_3 = \dfrac{1}{2i} \begin{pmatrix} 1 & 0 \\ 0 & -1 \end{pmatrix}. \end{cases}$$

## C. Infinitesimal Rotations

If we apply a rotation $R$ to a pair of functions $(\psi_1, \psi_2)$, the result is given by (25.3). We can achieve the same result in two steps: First we apply the rotation to the point $P$, thus transforming every function $\psi_\mu$ into a function $\varphi_\mu$ defined by

$$\varphi_\mu(P') = \psi_\mu(P),$$

and next we apply a linear transformation with matrix $T(R)$ to the pair $(\varphi_1, \varphi_2)$:

(25.8)
$$\psi'_\mu(P') = \Sigma\, t_{\mu\nu} \varphi_\nu(P').$$

Now let us apply an infinitesimal rotation, e.g. around the $x$-axis. This means: we have to apply a rotation $R_\alpha$ over an angle $\alpha$ about the $x$-axis,

---

[3] A simpler proof of this theorem was given in footnote 2 of my paper in Math. Zeitschr. **36**, p. 781.

differentiate with respect to $\alpha$, and put $\alpha = 0$. If we differentiate the right side of (25.8) with respect to $\alpha$, we have to differentiate the products $t_{\mu\nu}\varphi_\nu$ in the usual way by first differentiating the second factor, next differentiating the first factor, and finally adding the two partial results. This means: we have first to apply an infinitesimal rotation to the point $P$, next to apply to the pair $(\psi_1, \psi_2)$ the linear transformation corresponding to the infinitesimal rotation in the representation $\varrho_{\frac{1}{2}}$ without changing the point $P$, and finally to add the two results.

From § 20 we know that the first step, the infinitesimal rotation of the point $P$, applied to the pair $(\psi_1, \psi_2)$ or $\Psi$, yields the result

$$\left(\frac{\partial \Psi}{\partial \alpha}\right)_{\alpha=0} = -\left(y\frac{\partial}{\partial z} - z\frac{\partial}{\partial y}\right)\Psi \ .$$

The second step yields $I_x \Psi$, where $I_x$ is the two-row matrix defined in (25.7). Hence the total result of an infinitesimal rotation applied to $\Psi$ is

$$K_x\Psi = -\left(y\frac{\partial}{\partial z} - z\frac{\partial}{\partial y}\right)\Psi + \frac{1}{2i}s_1\Psi \ .$$

Just so, we have

$$K_y\Psi = -\left(z\frac{\partial}{\partial x} - x\frac{\partial}{\partial z}\right)\Psi + \frac{1}{2i}s_2\Psi$$

$$K_z\Psi = -\left(x\frac{\partial}{\partial y} - y\frac{\partial}{\partial x}\right)\Psi + \frac{1}{2i}s_3\Psi \ .$$

## D. The Angular Momenta

We now want to pass from the infinitesimal rotations to the angular momenta. In quantum mechanics without spin, the angular momentum with respect to the x-axis is

(25.9)
$$\hbar L_x = -i\hbar\left(y\frac{\partial}{\partial z} - z\frac{\partial}{\partial y}\right).$$

Hence, if we multiply the operator of the total infinitesimal rotation

(25.10)
$$K_x = -\left(y\frac{\partial}{\partial z} - z\frac{\partial}{\partial y}\right) + \frac{1}{2i}s_1$$

by $i\hbar$, the first term gives just the right orbital angular momentum operator $\hbar L_x$. Thus it seems reasonable to expect that the second term, multiplied by the same factor $i\hbar$, would give the spin momentum.

A convincing proof that this is correct was given by Dirac in his famous paper of 1928: "The Quantum Theory of the Electron", Proc. Royal Soc. A **117**, p. 610. The proof is based upon the theorem of the conservation of momentum, which Dirac assumes to be correct. In fact all measurements of the angular momenta of electrons are based upon the assumption of conservation of the total angular momentum. The idea of the proof is as follows.

Let us suppose an electron to move in a field with rotational symmetry about the $x$-axis. In this case, the Hamiltonian $H$ commutes with the operator of a rotation $R$ about the $x$-axis:

$$RH = HR$$

Differentiating $R$ with respect to the angle of rotation $\alpha$, and putting $\alpha = 0$, we obtain the operator of an infinitesimal rotation $K_x$. Hence

$$K_x H = H K_x .$$

This means that for $K_x$ a conservation law holds. Now $K_x$ is a sum of two terms

$$K_x = -\left(y\frac{\partial}{\partial x} - z\frac{\partial}{\partial y}\right) + \frac{1}{2i}s_1 .$$

The first term, multiplied by $\hbar i$, is the orbital momentum operator. Hence

$$(25.11) \qquad\qquad \hbar i \cdot \frac{1}{2i} s_1 = \frac{1}{2}\hbar s_1$$

is just the operator we have to add to the orbital momentum operator to find a total momentum operator for which the law of conservation holds. It follows that

$$\tfrac{1}{2}\hbar s_1$$

is just the spin momentum in the $x$-direction. Just so, $\frac{1}{2}\hbar s_2$ and $\frac{1}{2}\hbar s_3$ are the spin momenta in the $y$- and $z$-direction.

The states $(\psi_1, 0)$ and $(0, \psi_2)$ are eigenfunctions of the operator $s_1$ with eigenvalues $+1$ and $-1$. Hence the spin momentum has in these states the value $+\frac{1}{2}\hbar$ and $-\frac{1}{2}\hbar$, as it should be.

## E. The Doublet Splitting of the Alkali Terms

In what follows, the pair $(1, 0)$ will be denoted by $u_1$ and the pair $(0, 1)$ by $u_2$. Any pair $\Psi = (\psi_1, \psi_2)$ can now be written as

$$(25.12) \qquad \Psi = \psi_1 u_1 + \psi_2 u_2,$$

$\psi_1$ and $\psi_2$ being functions of the space coordinates $q$.

From (25.12), we can immediately deduce the Doublet Splitting of the Alkali Terms. In fact, as long as the spin-orbit interaction is neglected, $\psi_1$ and $\psi_2$ are solutions of one and the same differential equation. For any pair of quantum numbers $(n, l)$, we have $2l + 1$ linearly independent solutions of this differential equation:

$$\psi_l^{(m)} \qquad (m = l, l-1, \ldots, -l).$$

If we multiply these $2l + 1$ solutions by $u_1$ or $u_2$, we obtain $2(2l + 1)$ products

$$\psi_l^{(m)} u_1 \quad \text{and} \quad \psi_l^{(m)} u_2$$

which are transformed under rotations according to the representation

$$\varrho_l \times \varrho_{\frac{1}{2}} = \varrho_{l+\frac{1}{2}} + \varrho_{l-\frac{1}{2}} \quad (\text{for} \quad l > 0)$$

or

$$= \varrho_{\frac{1}{2}} \qquad (\text{for} \quad l = 0).$$

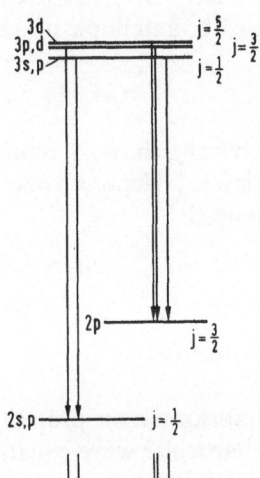

Fig. 6. Fine-structure of the line $H_\alpha$

Assuming the law of spin-orbit interaction to be invariant under rotations, we may conclude that only the terms $\varrho_{l+\frac{1}{2}}$ and $\varrho_{l-\frac{1}{2}}$ can separate, a further separation being impossible. Thus one obtains a doublet for $l > 0$, a singlet for $l = 0$. The quantum number

$$j = l + \tfrac{1}{2} \quad \text{or} \quad j = l - \tfrac{1}{2}$$

characterizing the representation $\varrho_j$ is called the *inner quantum number* of the electron. In any of the two cases, we have just $2j + 1$ eigenfunctions having exactly the same energy. If a magnetic field disturbs the rotational symmetry, the degeneration is removed and one obtains $2j + 1$ separate terms in each case.

### F. The Inversion $s$

We now investigate how the spin functions $u_1$ and $u_2$ behave under the inversion $s$:

$$x' = -x, \quad y' = -y, \quad z' = -z.$$

Once more, we may assume that $u_1$ and $u_2$ are transformed under $s$ into linear combinations of $u_1$ and $u_2$:

$$su_1 = u_1 s_{11} + u_2 s_{21},$$

$$su_2 = u_1 s_{12} + u_2 s_{22}.$$

Since the inversion $s$ commutes with all rotations $R$, the matrix $S = (s_{ij})$ must commute with all matrices of the irreducible representation $\varrho_{\frac{1}{2}}$. By Schur's Lemma, $S$ must be a multiple of the identity matrix

$$S = \lambda 1.$$

The value of $\lambda$ is quite arbitrary, since $\lambda \Psi$ represents the same state as $\Psi$. The simplest choice of $\lambda$ is $\lambda = 1$. Hence we may assume that the inversion $s$ leaves $u_1$ and $u_2$ unchanged:

$$su_1 = u_1, \quad su_2 = u_2.$$

## § 26. Dirac's Wave Equation

The derivation of the transformation properties of the spin functions was independent of any particular wave equation. This derivation has a compelling character, and the results can also be used in the case of several electrons.

For a single electron, P.A.M. Dirac[4] has set up a wave equation having the following properties:

1) it is invariant under Lorentz transformations,

2) it gives exactly the right doublet splitting of the hydrogen terms,

3) it gives exactly the right value for the magnetic moment of the electron.

Dirac's starting point is the observation that Schrödinger's time-dependent wave equation

$$i\hbar\, \partial_t \Psi = H\Psi \qquad \left(\partial_t = \frac{\partial}{\partial t}\right)$$

contains only the first derivative of $\Psi$ with respect to time, which implies that the state of the system at any time $t$ is completely determined by the initial state at time $t = 0$. Dirac expresses this by saying that the probability of any dynamical variable at any time having a value lying between specific limits is known as soon as the initial wave function $\Psi$ is known.

Dirac now set himself the problem, to find a first-order differential equation for $\Psi$, invariant under Lorentz transformations and equivalent to Gordon's second order equation in the case of no field.

In what follows, we shall use the following notation:

$$x^0 = ct, \qquad x^1 = x, \qquad x^2 = y, \qquad x^3 = z$$

$$\partial_0 = \frac{1}{c}\partial_t, \qquad \partial_1 = \partial_x, \qquad \partial_2 = \partial_y, \qquad \partial_3 = \partial_z$$

$$p_\alpha = -i\hbar\, \partial_\alpha \qquad (\alpha = 0, 1, 2, 3)$$

$$p^0 = -p_0, \qquad p^k = p_k \qquad (k = 1, 2, 3)$$

$$p_4 = p^4 = -ip^0 = +ip_0 .$$

In the case of no field, Gordon's second order wave equation can be written as

(26.1) $$\{-(p^0)^2 + (p^1)^2 + (p^2)^2 + (p^3)^2 + \mu c^2\}\, \Psi = 0$$

or as

(26.2) $$\left\{\sum_1^4 (p^\alpha)^2 + \mu c^2\right\} \Psi = 0 .$$

---

[4] P. A. M. Dirac: The Quantum Theory of the Electron. Proc. Royal Soc. (A) 117, p. 610, and 118, p. 351 (1928).

Dirac now proposes to replace (26.2) by a first-order wave equation

(26.3)
$$\left(i\sum_1^4 \gamma_\alpha p^\alpha + \mu c\right)\Psi = 0$$

in which the $\gamma_\alpha$ are supposed to be matrices satisfying the conditions

(26.4)
$$\gamma_\alpha \gamma_\beta + \gamma_\beta \gamma_\alpha = 2\delta_{\alpha\beta} \begin{cases} = 2 & (\alpha = \beta) \\ = 0 & (\alpha \neq \beta). \end{cases}$$

In fact, if these conditions are satisfied, one can write (26.3) as

$$(\Sigma \gamma_\alpha p^\alpha)\Psi = i\mu c\Psi$$

which implies

$$(\Sigma \gamma_\alpha p^\alpha)^2 \Psi = -\mu^2 c^2 \Psi$$

Carrying out the multiplications on the left, making use of (26.4), one obtains

$$\Sigma (p^\alpha)^2 \Psi = -\mu^2 c^2 \Psi$$

which is equivalent to (26.2). Hence in the case of no field Gordon's Eq. (26.1) is a consequence of Dirac's Eq. (26.3).

Dirac now shows that the conditions (26.4) can be satisfied by matrices having four rows and columns. Dirac's matrices are constructed by means of Pauli's matrices

$$s_1 = \begin{pmatrix} 0 & 1 \\ 1 & 0 \end{pmatrix}, \quad s_2 = \begin{pmatrix} 0 & -i \\ i & 0 \end{pmatrix}, \quad s_3 = \begin{pmatrix} 1 & 0 \\ 0 & -1 \end{pmatrix}$$

as follows:

$$\gamma_1 = \begin{pmatrix} 0 & s_1 \\ s_1 & 0 \end{pmatrix} = \begin{pmatrix} 0 & 0 & 0 & 1 \\ 0 & 0 & 1 & 0 \\ 0 & 1 & 0 & 0 \\ 1 & 0 & 0 & 0 \end{pmatrix}$$

$$\gamma_2 = \begin{pmatrix} 0 & s_2 \\ s_2 & 0 \end{pmatrix} = \begin{pmatrix} 0 & 0 & 0 & -i \\ 0 & 0 & i & 0 \\ 0 & -i & 0 & 0 \\ i & 0 & 0 & 0 \end{pmatrix}$$

$$\gamma_3 = \begin{pmatrix} 0 & s_3 \\ s_3 & 0 \end{pmatrix} = \begin{pmatrix} 0 & 0 & 1 & 0 \\ 0 & 0 & 0 & -1 \\ 1 & 0 & 0 & 0 \\ 0 & -1 & 0 & 0 \end{pmatrix}.$$

$$\gamma_4 = \begin{pmatrix} 1 & 0 \\ 0 & -1 \end{pmatrix} = \begin{pmatrix} 1 & 0 & 0 & 0 \\ 0 & 1 & 0 & 0 \\ 0 & 0 & -1 & 0 \\ 0 & 0 & 0 & -1 \end{pmatrix}$$

Thus, Dirac obtains a wave Eq. (26.3), in which the wave function $\Psi$ has four components arranged in one column:

$$\Psi = \begin{pmatrix} \psi_1 \\ \psi_2 \\ \psi_3 \\ \psi_4 \end{pmatrix}.$$

Instead of using Dirac's matrices $\gamma_\alpha$, one can use any other set of matrices $\gamma'_\alpha$ obtained from $\gamma_\alpha$ by a transformation with an arbitrary matrix $\tau$:

(26.5) $$\gamma'_\alpha = \tau \gamma_\alpha \tau^{-1}.$$

The resulting equation

(26.6) $$\left( i \sum_1^4 \gamma'_\alpha p^\alpha + \mu c \right) \Psi' = 0$$

is equivalent to (26.3), for if one puts $\Psi' = \tau \Psi$ and substitutes (26.5) in (26.6), one obtains

$$(i \sum \tau \gamma_\alpha \tau^{-1} p^\alpha + \mu c) \tau \Psi = 0$$

or

$$\tau (i \sum \gamma_\alpha p^\alpha + \mu c) \Psi = 0$$

which is equivalent to (26.3).

A useful set of matrices $\gamma'_\alpha$, upon which Flügge's treatment of the Dirac equation[5] is based, is obtained from Dirac's set by choosing

$$\tau = \begin{pmatrix} 1 & 0 \\ 0 & i \end{pmatrix} = \begin{pmatrix} 1 & 0 & 0 & 0 \\ 0 & 1 & 0 & 0 \\ 0 & 0 & i & 0 \\ 0 & 0 & 0 & i \end{pmatrix}.$$

The resulting matrices are

$$\gamma'_k = \begin{pmatrix} 0 & -i\sigma_k \\ i\sigma_k & 0 \end{pmatrix} \quad (k=1,2,3)$$

$$\gamma'_4 = \begin{pmatrix} 1 & 0 \\ 0 & -1 \end{pmatrix}.$$

In what follows, we shall use this *standard set* of matrices $\gamma_\alpha$. Flügge calls it the *standard representation* and uses it consistently in his Chapter VI.

If an electromagnetic field is present, derived from an electrical potential $A^0$ and a magnetic vector potential $(A^1, A^2, A^3)$, one has to replace the operators $p^0, p^k, p^4$ by

$$D^0 = p^0 + \frac{e}{c} A^0$$

$$D^k = p^k + \frac{e}{c} A^k \quad (k=1,2,3)$$

$$D^4 = -iD^0,$$

thus obtaining a Wave Equation

(26.7)
$$\left( i \sum_1^4 \gamma'_\alpha D^\alpha + \mu c \right) \Psi = 0,$$

which is invariant under Lorentz transformations.

From now on, the standard matrices $\gamma'_\alpha$ will be denoted by $\gamma_\alpha$, as in Flügge's book quoted before.

---

[5] S. Flügge: Practical Quantum Mechanics II, Chapter VI. Berlin-Heidelberg-New York: Springer 1971.

A generalization of Dirac's matrices to $n$ dimensions was given by Brauer and Weyl[6]. I shall give an outline of their theory.

Consider an arbitrary non-singular quadratic form in $n$ variables

$$(26.8) \qquad\qquad Q = \Sigma \, g_{\alpha\beta} \, x^\alpha x^\beta$$

with complex coefficients $g_{\alpha\beta} = g_{\beta\alpha}$. "Non-singular" means: the determinant of the $g_{\alpha\beta}$ is not zero. The "Clifford Algebra" of the form $Q$ is an algebra over the complex number field generated by a unit element 1 and $n$ elements

$$u_1, \ldots, u_n$$

subject to the conditions

$$(26.9) \qquad\qquad u_\alpha u_\beta + u_\beta u_\alpha = 2g_{\alpha\beta} \cdot 1 \, .$$

The algebra has a basis consisting of the $2^n$ elements

$$1, u_\alpha, u_\alpha u_\beta, u_\alpha u_\beta u_\gamma, \ldots (\alpha < \beta < \gamma \ldots) \, .$$

The following theorems hold:

1. The Clifford Algebra is isomorphic to a full matrix algebra if $n$ is even, and to a direct sum of two full matrix algebras if $n$ is odd.

2. If $n$ is even, all automorphisms of the algebra are inner automorphisms, i.e. they are given by

$$(26.10) \qquad\qquad u'_\alpha = \tau u_\alpha \tau^{-1}$$

In the field of complex numbers, every non-singular quadratic form $Q$ can be reduced, by a linear transformation of coordinates, to the special form

$$Q = \Sigma \, x^\alpha x^\alpha$$

with coefficients $\delta_{\alpha\beta}$. The linear transformations leaving invariant $Q$ are orthogonal transformations. If such a transformation $T$ is applied to the elements $u_\alpha$ of the Clifford algebra, one obtains a new set $u'_\alpha$ satisfying the same conditions (26.9), and hence an isomorphism of the algebra. If $n$ is even, one can apply Theorem 2 and conclude that the orthogonal transformation $T$ can be obtained by transforming the $u_\alpha$ according to formula (26.10). Thus, to any orthogonal transformation $T$ corresponds an element $\tau$ of the Clifford algebra, i.e. (since the Clifford algebra is a full matrix algebra) a matrix $\tau$ of degree $2^m$ with $n = 2m$. If $n$ is odd, one can obtain a similar result by using the "second Clifford algebra", which is the sub-algebra generated by the even products

$$1, u_\alpha u_\beta, u_\alpha u_\beta u_\gamma u_\delta, \ldots$$

which is, for $n = 2m + 1$, a full matrix algebra of order $2^m$.

For further details see the paper of Brauer and Weyl quoted before, or the book of C. Chevalley: Algebraic Theory of Spinors.

# § 27. Two-Component Spinors

## A. Dirac's Equation Rewritten

In §23 we have determined all differentiable representations of the restricted Lorentz group. We may also express this by saying that we have determined all kinds of quantities that are linearly transformed

---

[6] R. Brauer and H. Weyl: Spinors in $n$ dimensions. American J. of Math. **57**, p. 245.

under the Lorentz group. We have seen that all these quantities can be written as "spinors" or "spin-tensors" with dotted and undotted indices

(27.1)
$$c_{\lambda\mu\dots\nu,\varrho\cdot\sigma\cdot\dots\tau\cdot}$$

which are symmetric in the dotted as well as in the undotted indices. Every undotted index takes the values 1 and 2, and every dotted index takes the values $1\cdot$ and $2\cdot$. The tensor components (27.1) are transformed just like products

$$a_\lambda a_\mu \dots a_\nu b_{\varrho\cdot} b_{\sigma\cdot} \dots b_{\tau\cdot},$$

in which the pairs $\begin{pmatrix} a_1 \\ a_2 \end{pmatrix}$ are transformed like vectors under the group SL(2), whereas the pairs $\begin{pmatrix} b_{1\cdot} \\ b_{2\cdot} \end{pmatrix}$ are transformed by the complex conjugate linear transformations.

Now Dirac's four-component spinors are quantities that are linearly transformed under the Lorentz Group. Hence it must be possible to find expressions for the four-component spinors in terms of two-component spinors, and to re-write Dirac's equation in the notation of Spinor Analysis. This can be done as follows.

Following Flügge, we may write the four-component spinor $\Psi$ as

$$\Psi = \begin{pmatrix} \psi_a \\ \psi_b \end{pmatrix} \quad \text{with} \quad \psi_a = \begin{pmatrix} \psi_1 \\ \psi_2 \end{pmatrix} \quad \text{and} \quad \psi_b = \begin{pmatrix} \psi_3 \\ \psi_4 \end{pmatrix}.$$

Using the standard set of matrices

$$\gamma_k = \begin{pmatrix} 0 & -is_k \\ is_k & 0 \end{pmatrix}, \quad \gamma_4 = \begin{pmatrix} 1 & 0 \\ 0 & -1 \end{pmatrix}$$

we may write the Wave Eq. (26.7) as

(27.2)
$$\sum_1^3 s_k D^k \psi_b + (iD^4 + \mu c)\psi_a = 0$$

$$-\sum_1^3 s_k D^k \psi_a + (-iD^4 + \mu c)\psi_b = 0$$

or, introducing $D^0 = iD^4$, as

(27.3)
$$\begin{cases} \sum_1^3 s_k D^k \psi_b + (D^0 + \mu c)\psi_a = 0 \\ \sum_1^3 s_k D^k \psi_a + (D^0 - \mu c)\psi_b = 0. \end{cases}$$

In the case of no field, the operators $D^0$ and $D^k$ are

$$D^0 = p^0 = -p_0 = \quad i\hbar\,\partial_0 \quad \left(\partial_0 = \frac{1}{c}\,\partial_t\right)$$

$$D^k = p^k = \quad p_k = -i\hbar\,\partial_k$$

hence (27.3) can be written as

(27.4)
$$\begin{cases} -i\hbar \sum_1^3 s_k \partial_k \psi_b + (i\hbar\,\partial_0 + \mu c)\psi_a = 0 \\[2mm] -i\hbar \sum_1^3 s_k \partial_k \psi_a + (i\hbar\,\partial_0 - \mu c)\psi_b = 0. \end{cases}$$

It is easy to find solutions of these equations in the form of plane waves

(27.5)
$$\psi_a = C_a e^{i(k_1 x + k_2 y + k_3 z - \omega t)}$$
$$\psi_b = C_b e^{i(k_1 x + k_2 y + k_3 z - \omega t)}.$$

Substituting (27.5) into (27.4), one finds two kinds of solutions:

1) solutions with positive energy $E = \hbar\omega$, in which $\psi_b$ is very small as compared with $\psi_a$;

2) solutions with negative energy, in which $\psi_a$ is very small as compared with $\psi_b$.

Type 1) corresponds to electrons, type 2) to positrons. More precisely one has to assume, according to Dirac, that all states of negative energy are occupied with the exception of a few holes, which appear as positrons, whereas only a few states of positive energy are occupied by electrons.

We now introduce spinors $\psi^{\mu'}$ and $\chi_\nu$, putting

$$\psi_a + \psi_b = \psi^{\cdot} = \begin{pmatrix} \psi^{1'} \\ \psi^{2'} \end{pmatrix}$$

$$\psi_a - \psi_b = \chi = \begin{pmatrix} \chi_1 \\ \chi_2 \end{pmatrix}.$$

Adding and subtracting the Eqs. (27.3), one obtains

(27.6)
$$\begin{cases} i\hbar\left(-\partial_0 + \sum_1^3 s_k \partial_k\right)\psi^{\cdot} = \mu c \chi \\[2mm] i\hbar\left(-\partial_0 - \sum_1^3 s_k \partial_k\right)\chi = \mu c \psi. \end{cases}$$

Denoting the unit matrix by $s_0$ and putting

$$\partial^0 = -\partial_0, \qquad \partial^k = \partial_k$$

we may rewrite (27.6) as

(27.7)
$$\begin{cases} i\hbar(s_0\partial^0 + \Sigma\, s_k\partial^k)\psi^{\cdot} = \mu c\chi \\ i\hbar(s_0\partial^0 - \Sigma\, s_k\partial^k)\chi = \mu c\psi^{\cdot}. \end{cases}$$

In § 23 C the elements of the matrices $s_0, s_1, s_2, s_3$ were denoted by $\sigma_{k\lambda\mu}$. Just so, the elements of the matrices $s_0, -s_1, -s_2, -s_3$ were denoted by $\sigma_k'^{\mu\cdot\nu}$. In this notation, the Eqs. (27.7) may be written as

(27.8)
$$\begin{cases} i\hbar\, \Sigma\, \sigma_{k\lambda\mu}\cdot \partial^k \psi^{\mu^{\cdot}} = \mu c\chi_\lambda \\ i\hbar\, \Sigma\, \sigma_k'^{\mu\cdot\nu}\, \partial^k \chi_\nu = \mu c\psi^{\mu^{\cdot}} \end{cases}$$

(summation over upper and lower indices, as usual).

## B. Weyl's Equation

If the mass $\mu$ is zero, (27.8) yields two separate wave equations, one for $\psi^{\cdot}$ and one for $\chi$:

(27.9)
$$\Sigma\, \sigma_{k\lambda\mu}\cdot \partial^k \psi^{\mu^{\cdot}} = 0\,,$$

(27.10)
$$\Sigma\, \sigma_k'^{\mu\cdot\nu}\, \partial^k \chi_\nu = 0\,,$$

or, written in the simpler notation used in (27.6)

(27.11)
$$(-\partial_0 + \Sigma\, s_k\partial_k)\psi^{\cdot} = 0\,,$$

(27.12)
$$(-\partial_0 - \Sigma\, s_k\partial_k)\chi = 0\,.$$

These wave equations for particles of mass zero were first proposed by H. Weyl. They are not invariant with respect to spatial reflections. For this and other reasons Pauli rejected Weyl's equations. As far as electrons were concerned, this rejection was fully justified, but in 1957 Lee and Yang[7] proposed to apply one of Weyl's equations to neutrinos. Experiments made by C. S. Wu and others have shown that in $\beta$-decay processes, in which a neutrino is emitted, parity is not conserved. Lee and Yang proposed to use Eq. (27.12) but Feynman and Gell-Mann

---

[7] T. D. Lee and C. N. Yang: Parity Non-Conservation and a Two-Component Theory of the Neutrino. Phys. Revue **105**, p. 1671 (1957). See also: R. P. Feynman and M. Gell-Mann: Theory of Fermi-Interaction, Phys. Rev. **109**, p. 193.

showed that Eq. (27.11) is in better accordance with experimental evidence, or, as they put it, that "neutrinos spin to the left".

To explain what this means, let us solve (27.11) by a plane wave

$$(27.13) \qquad \psi^{\cdot} = a^{\cdot} e^{i(kx - \omega t)} \qquad [kx = k_1 x^1 + k_2 x^2 + k_3 x^3]$$

$a^{\cdot}$ being a constant two-component spinor. If $\psi^{\cdot}$ is a solution of (27.11), $a^{\cdot}$ has to satisfy the condition

$$(27.14) \qquad (\omega + \Sigma s_r k_r) a^{\cdot} = 0.$$

The momentum vector of this plane wave is $\hbar k$. To simplify the calculations we may assume that this vector is directed along the positive $z$-axis. Equation (27.14) now simplifies to

$$(27.15) \qquad (\omega + s_3 k_3) a^{\cdot} = 0.$$

The eigenvalues of the matrix $s_3$ are $+1$ and $-1$, hence the eigenvalues of $\omega + s_3 k_3$ are

$$\omega + k_3 \quad \text{and} \quad \omega - k_3.$$

The Eq. (27.15) implies that 0 is an eigenvalue. Since $\omega$ and $k_3$ are positive, $\omega + k_3$ cannot be zero. So the only possibility is

$$\omega - k_3 = 0, \quad \text{hence} \quad k_3 = \omega.$$

The spin vector $a^{\cdot}$ is an eigenvector belonging to the eigenvalue $-1$ of $s_3$, that is,

$$a^{\cdot} = \begin{pmatrix} a^{1^{\cdot}} \\ a^{2^{\cdot}} \end{pmatrix} = \begin{pmatrix} 0 \\ 1 \end{pmatrix}.$$

This means: if the velocity of the plane wave is directed upwards, the spin is directed downwards, which can indeed be expressed by saying that "neutrinos spin to the left".

For the history of Neutrino theory see the contribution of C. S. Wu in the Pauli Memorial Volume, p. 249—303, New York: Interscience 1960.

## § 28. *The Several Electron Problem. Multiplet Structure. Zeeman Effect*

We now return to the non-relativistic theory. The state of a system of $f$ electrons is given by a function

$$\psi(q_1, q_2, ..., q_f; \sigma_1, ..., \sigma_f)$$

where the $q_m$ are spatial coordinates and $\sigma_m$ is the spin-coordinate $\sigma_z$ of the $m$-th electron. Introducing the basis vectors $u_1, u_2$ in the spin space of the first electron as in § 25, and likewise $v_1, v_2$ for the second electron, etc. we can also write our function as

$$(28.1) \qquad \psi(q_1, ..., q_f; \sigma_1, ..., \sigma_f) = \Sigma \, \psi_{\lambda\mu...\nu} u_\lambda v_\mu ... \, w_\nu.$$

Under rotations of the space, the functions (28.1) are transformed in such a way that each pair of basis vectors is transformed like $u_1, u_2$ according to the representation $\varrho_{\frac{1}{2}}$ of the rotation group, whereas the $\psi_{\lambda\mu...\nu}$ are transformed like normal spatial functions. The products $u_\lambda v_\mu ... \, w_\nu$ are consequently transformed according to the representation $\varrho_{\frac{1}{2}} \times \varrho_{\frac{1}{2}} \times \cdots \times \varrho_{\frac{1}{2}}$. The $u_\lambda, v_\mu, ...$ are left invariant by the reflection $s$.

If one has a system of eigenfunctions

$$\psi^{(1)}(q), ..., \psi^{(k)}(q)$$

of the spin-free Schrödinger equation corresponding to the eigenvalue $E$ of the Hamiltonian, the $k \cdot 2^f$ products

$$(28.2) \qquad \psi^{(\alpha)} u_\lambda v_\mu ... \, w_\nu$$

also satisfy the Schrödinger equation, as long as spin perturbation terms are neglected. In order to see how this $(k \cdot 2^f)$-tuple term is split up by the spin effect, we first investigate how it is transformed under rotations. Let the $\psi^{(\alpha)}$ be transformed according to the irreducible representation $\varrho_L$. Then the products (28.2) are transformed according to the representation

$$(28.3) \qquad \varrho_L \times \varrho_{\frac{1}{2}} \times \varrho_{\frac{1}{2}} \times \cdots \times \varrho_{\frac{1}{2}}.$$

After reduction of this representation one obtains the irreducible subspaces which can subsequently be separated by the spin perturbation.

It is useful to carry out the reduction of the representation (28.3) in such a way that the factors $\varrho_{\frac{1}{2}}$ are first multiplied together:

$$\varrho_{\frac{1}{2}} \times \varrho_{\frac{1}{2}} = \varrho_0 + \varrho_1$$

$$\varrho_{\frac{1}{2}} \times \varrho_{\frac{1}{2}} \times \varrho_{\frac{1}{2}} = \varrho_{\frac{1}{2}} + \varrho_{\frac{1}{2}} + \varrho_{\frac{1}{2}}.$$

After this, each individual term $\varrho_S$ so obtained may be multiplied by $\varrho_L$ according to the equation

$$(28.4) \qquad \varrho_L \times \varrho_S = \Sigma \, \varrho_J \qquad (J = L + S, ..., |L - S|).$$

Hence, in the vector scheme the spins $\frac{1}{2}\hbar$ of the individual electrons are first combined to form a resultant $\hbar S$, which is then combined with the orbital angular momentum $L$ to form a resultant of length $\hbar J$, whose component in the $z$-direction can again assume the values $\hbar M$ ($M = J, J-1, \ldots, -J$).

The number $J$ is called the *azimuth quantum number*, $S$ the *spin number*, $J$ the *inner quantum number* and $M$ the *magnetic quantum number*. $S$ and $J$ are integers for even numbers of electrons, otherwise half-integers.

The various terms $\varrho_J$ resulting from a product (28.4) by means of reduction and spin perturbation are combined into a *multiplet*. The multiplet is called *normal* if the terms with largest $J$ are located highest, otherwise it is called *inverted*.

It will be shown in the next chapter that of the various theoretically possible values of $S$ resulting from the multiplication $\varrho_{\frac{1}{2}} \times \varrho_{\frac{1}{2}} \times \cdots$ only one actually occurs in nature. This one value of $S$ gives rise to a complete multiplet (28.4), for which all theoretically possible $J$-values actually occur.

If $L \geq S$, then the number of terms of a multiplet is $2S+1$, but if $L < S$, the multiplicity "cannot fully develop": only $2L+1$ terms occur; in particular, in the case $L = 0$ ($S$-term), only one term occurs (singlet). In spite of this, one *always* speaks of doublet terms in the case $S = \frac{1}{2}$, *always* of triplet-terms for $S = 1$, etc. Thus, we have

$$(28.5) \quad \begin{cases} \text{singlet terms} \quad {}^1S, {}^1P, {}^1D, \ldots (S=0) \\ \text{doublet terms} \quad {}^2S, {}^2P, {}^2D, \ldots (S=\tfrac{1}{2}) \\ \text{triplet terms} \quad {}^3S, {}^3P, {}^3D, \ldots (S=1) \\ \text{etc.} \end{cases}$$

The symbol ${}^2P$ is pronounced as "doublet $P$". The reason for this terminology lies in the selection rule for $S$, which will now be established. The components of a multiplet will be distinguished by means of an index $J$ affixed on the lower right. For example, a ${}^3P$-term consists of the components ${}^3P_0, {}^3P_1, {}^3P_2$.

The behaviour of the eigenfunctions (28.2) at the origin under the inversion $s$ is quite easy to determine, because the $u_\lambda$ etc. remain invariant. Thus, if the $\psi^{(\alpha)}$ correspond to the reflection character

$$w = (-1)^{l_1 + \cdots + l_f}$$

then the products (28.4) also correspond to the same reflection character, so that nothing changes after introduction of the spin perturbation.

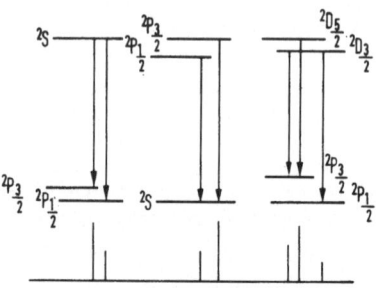

Fig. 7. Transitions between normal doublets

The following *exact selection rules* hold:

(28.6)
$$\begin{cases} J \to J-1, J, J+1 \quad \text{(but not } 0 \to 0) \\ M \to M-1, M, M+1 \\ w \to -w. \end{cases}$$

The same additional results on the intensity and polarization of the emitted light hold that we derived in § 22. For the proof one need only replace $L$ by $J$ everywhere in § 22. Indeed, the proof rested solely on the properties of the representation $\varrho_L$, which is now replaced by $\varrho_J$.

The selection rule for $J$ tells us which transitions between the terms of any two multiplets are possible. In Fig. 7 the admissible inter-combinations along with the positions and intensities of the lines are shown for some types of doublet terms.

The selection rule for $w$ is precisely the rule of Laporte (see § 21). The selection rule for $M$ holds when an axially symmetric perturbation is present which removes the $(2J+1)$-tuple rotation degeneracy (Zeeman or Stark effect). The relationships of the intensities of the split-up lines emitted under such a small perturbation can be taken from (21.9).

As long as the multiplet split-up (spin effect) is small (hence, in particular, for the lighter elements) the following selection rules hold:

(28.7)
$$\begin{cases} L \to L-1, L, L+1 \quad \text{(but not } 0 \to 0) \\ S \to S. \end{cases}$$

In fact, if one multiplies the approximating eigenfunctions (28.2) by $\Sigma x, \Sigma y$ or $\Sigma z$ and expands in terms of the same functions, then (28.7) results simply from the fact that the products $u_\lambda v_\mu \ldots$ are left unchanged and the $(\Sigma x)\psi^{(\alpha)}$ etc. are expanded in terms of the $\psi^{(\beta)}$. Hence, the same terms $\psi^{(\beta)}$ occur which did so without spin perturbation and these must therefore satisfy the old selection rule for $L$, while the spin functions

$u_\lambda v_\mu \ldots$ as well as their linear combinations belonging to the representation $\varrho_S$ remain completely unchanged in the expansion.

Because of the effect of the spin perturbation, lines can be emitted which violate the rules (28.7). For example, combinations between triplet and singlet terms are quite frequent among the heavy elements.

The rule $S \to S$ says that the entire series spectrum of an element decomposes into different systems of lines each corresponding to a system of terms with one $S$-value. These term systems are called singlet, doublet, ... systems according to the scheme (28.5).

**Example.** For the lighter atoms with two valence electrons such as He, Be, Mg, we find one singlet and one triplet system which do not combine with one another (see Fig. 9 in § 29). The $S$-terms in both systems are singlets, although one speaks of a $^3S$ (pronounce: triplet $S$) term, because it belongs to the triplet system.

*The Anomalous Zeeman Effect.* According to § 25, the disturbance term in the wave equation which is linear in the magnetic field strength is given for a homogeneous field in the $z$-direction by

$$\kappa \mathcal{H} \cdot (\mathcal{L} + 2\mathcal{S}) = \kappa \mathcal{H} \cdot (\mathcal{M} + \mathcal{S}) = \kappa H_z (M_z + S_z).$$

Let us assume, to begin with, that the perturbation is small as compared with the multiplet split-up (weak magnetic field). Then, according to perturbation theory, for a linear space $\mathscr{V}_{2J+1}$ belonging to a multiplet line, we must form the expression $(M_z + S_z) \psi_J^{(M)}$, expand it in terms of the $\psi_J^{(M')}$, and then pick out the terms $\psi_J^{(M')}$ belonging to the *same* linear space $\mathscr{V}_{2J+1}$. Since $M_z \psi_J^{(M)} = M \psi_J^{(M)}$, we need only evaluate $S_z \psi_J^{(M)}$. Just so, we may evaluate $S_x \psi_J^{(M)}$ and $S_y \psi_J^{(M)}$. Thus we must expand in terms of the $\psi_J^{(M')}$ a set of functions transformed according to $\varrho_1 \times \varrho_J$. By the results of § 21, all expansion coefficients corresponding to a space $\mathscr{V}_{2J+1}$ are uniquely determined by purely group-theoretic considerations. Thus, if $S_x'$, $S_y'$, $S_z'$ are the operators arising from $S_x$, $S_y$, $S_z$ when one deletes all terms from the series expansion which do not lie in the space $\mathscr{V}_{2J+1}$, and if $M_x'$, $M_y'$, $M_z'$ are formed in an analogous way, then the $S_x'$, $S_y'$, $S_z'$ must necessarily coincide up to a factor $\beta$ with $M_x'$, $M_y'$, $M_z'$:

$$S_x' = \beta M_x', \quad S_y' = \beta M_y', \quad S_z' = \beta M_z'.$$

Hence,

$$(M_z' + S_z') \psi_J^{(M)} = (1 + \beta) M_z' \psi_J^{(M)} = (1 + \beta) M \psi_J^{(M)},$$

so that the $\psi_J^M$ are the first approximations to the eigenfunctions of the perturbed problem, and $(1 + \beta) M H_z$ is the magnetic split-up.

In order to determine the split-up factor $g = 1 + \beta$, we employ the following device: we form the scalar product

$$(\mathscr{S}'\mathscr{M}) = (\mathscr{M}'\mathscr{S}') = \beta \mathscr{M}'^2 = \beta J(J + 1).$$

Then

$$\mathscr{L}^2 = (\mathscr{M} - \mathscr{S}')^2 = \mathscr{M}^2 - \mathscr{M}\mathscr{S}' - \mathscr{S}'\mathscr{M} + \mathscr{S}'^2.$$

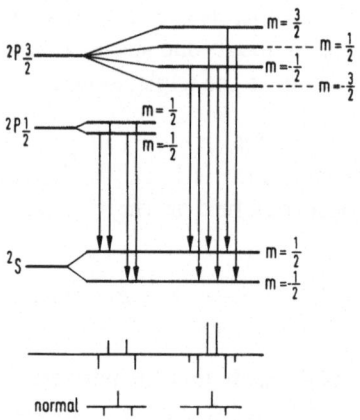

Fig. 8. Zeeman types $^2P \rightarrow {}^2S$.

Limiting consideration on the left and right in the last equation to those parts of the operators corresponding to the space $\mathscr{V}_{2J+1}$ and taking into consideration the fact that all lines of a multiplet correspond approximately to the eigenvalue $L(L + 1)$ of $\mathscr{L}^2$ and to the eigenvalue $S(S + 1)$ of $\mathscr{S}^2$, then we obtain (for small multiplet split-up):

$$L(L + 1) = J(J + 1) - 2\beta J(J + 1) + S(S + 1).$$

From this we calculate $\beta$ and obtain finally

$$g = 1 + \beta = 1 + \frac{J(J + 1) + S(S + 1) - L(L + 1)}{2J(J + 1)}.$$

This formula agrees with experience (see Landé's empirical formula (24.1) for $S = \frac{1}{2}$). Together with the selection rule $M \rightarrow M + 1, M, M - 1$ and the intensity rules, the formula determines the "typical" Zeeman split-up which reoccurs with each quantum jump $L \rightarrow L'$, $S \rightarrow S'$, $J \rightarrow J'$. Two examples of this situation are shown in Fig. 8. The lines polarized parallel to the magnetic field are directed upward in the figure and the remaining lines downward. For comparison purposes, a normal Zeeman effect is shown with the same scale in each case.

If the magnetic split-up is of the same order of magnitude as the multiplet split-up (stronger magnetic field), one must treat both perturbations together. See W. Heisenberg and P. Jordan: Anwendung der Quantenmechanik auf das Problem der anomalen Zeemaneffekte. Z. f. Physik **37**, p. 263 (1926).

Chapter V

# The Group of Permutations and the Exclusion Principle

## § 29. The Resonance of Equal Particles [1]

A stationary state of a pair of electrons, without spin and without interaction, is determined by an eigenfunction of the form

$$(29.1) \qquad \Psi(q_1, q_2) = \psi_1(q_1)\, \psi_2(q_2)$$

in which $\psi_1$ and $\psi_2$ are the normed eigenfunctions of the single electrons. If $E_1$ and $E_2$ are their energy values, the energy of the whole system is

$$E = E_1 + E_2 .$$

The same energy value belongs to another eigenfunction, which is obtained from (29.1) by a permutation (12) of the electrons:

$$(29.2) \qquad \Psi' = (12)\Psi = \psi_1(q_2)\, \psi_2(q_1) .$$

From the functions $\Psi$ and $\Psi'$ one can form the sum and difference

$$\Psi_s = \Psi + (12)\Psi ,$$
$$\Psi_a = \Psi - (12)\Psi .$$

The function $\Psi_s$ is *symmetric*, i.e.

$$(12)\Psi_s = \Psi_s$$

and the function $\Psi_a$ is *antisymmetric*, i.e.

$$(12)\Psi_a = -\Psi_a .$$

---

[1] See W. Heisenberg: Mehrkörperproblem und Resonanz in der Quantenmechanik. Zeitschr. f. Phys. **38**, p. 411 (1926).

The symmetric functions form a subspace of the Hilbert space, and the antisymmetric ones form another subspace. The two subspaces have only 0 in common. Every function $\Psi$ is the sum of a symmetric function

$$\tfrac{1}{2}(\Psi + (12)\Psi)$$

and an antisymmetric one

$$\tfrac{1}{2}(\Psi - (12)\Psi).$$

Hence the Hilbert space is a direct sum of the two subspaces of symmetric and antisymmetric functions. The two subspaces are orthogonal: all scalar products $\langle \Psi_s, \Psi_a \rangle$ are zero.

As long as the interaction is neglected, both states $\Psi_s$ and $\Psi_a$ have the same energy $E_1 + E_2$. We now consider the interaction as a perturbation. Since the interaction operator $W$ commutes with the permutation $(12)$, it transforms symmetric functions $\Psi_s$ into symmetric ones and antisymmetric functions $\Psi_a$ into antisymmetric ones, and we can apply Perturbation Theory to each of the two subspaces separately. The result of the perturbation will usually be a term split: the energy values belonging to the functions $\Psi_s$ and $\Psi_a$ will usually be different.

A rough estimate of the perturbed energy values may be obtained by applying *first order* perturbation theory. This means: the eigenfunctions (29.1) and (29.2) are retained, but the corrected eigenvalues are determined by means of a secular equation, as explained in § 7.

In the simplest case, the two electrons are in the same state: $\psi_1 = \psi_2$. To fix the ideas, let us consider the ground state of the Helium atom, in which both electrons are in the lowest s-state (1 s). In this case $\Psi_a$ is zero, and $\Psi_s$ is (but for a factor 2) just the product

$$\Psi = \psi_1(q_1)\,\psi_1(q_2).$$

Because $\psi_1$ is normed, $\Psi$ is normed as well:

$$\langle \Psi, \Psi \rangle = 1.$$

The Hamiltonian $H$ may be written as $H_0 + W$, where $H_0$ is the Hamiltonian without interaction, whereas $W$ is the Coulomb interaction energy $e^2/r_{12}$. The degree of the secular equation is 1, and the corrected energy value of the Hamiltonian $H = H_0 + W$ in the state $\Psi$ is:

$$
\begin{aligned}
E &= \langle \Psi, H\Psi \rangle \\
&= \langle \Psi, H_0\Psi \rangle + \langle \Psi, W\Psi \rangle \\
&= E_1 + E_2 + \int \psi_1^*(q_1)\,\psi_1^*(q_2)\,\frac{e^2}{r_{12}}\,\psi_1(q_1)\,\psi_1(q_2)\,dq_1\,dq_2 .
\end{aligned}
$$

Next consider the case in which the two electrons are in different states:

$$\psi_1 \neq \psi_2 .$$

To fix the ideas, let us suppose that $\psi_1$ is the eigenfunction of an $s$-state, whereas $\psi_2$ may be any one of the $(2l+1)$ eigenfunctions belonging to any value of $l$. For every fixed value of $m$, we may form the products $\Psi$ and $\Psi'$, and their sum $\Psi_s$ and difference $\Psi_a$, as before. Both products $\Psi$ and $\Psi'$ belong to one and the same value of the total quantum number $M$:

$$M = 0 + m = m .$$

Since the Hamiltonian $H = H_0 + W$ commutes with $L_z$, we may consider each value of $M$ separately, and we have a secular equation of degree 2 only. Since we may apply the operator $H$ to the symmetric and antisymmetric functions separately, we are left with two separate secular equations of degree 1. We just have to expand $H\Psi_s$, which is a symmetric function, into symmetric functions, and $H\Psi_a$ into antisymmetric functions:

(29.3)
$$\begin{cases} H\Psi_s = \alpha\Psi_s + \cdots \\ H\Psi_a = \beta\Psi_a + \cdots , \end{cases}$$

the terms $+\cdots$ being orthogonal to the main term $\alpha\Psi_s$ or $\beta\Psi_a$ respectively.

The coefficients $\alpha$ and $\beta$ on the right in (29.3) can be determined from the scalar products as follows:

(29.4)
$$\alpha = \frac{\langle \Psi_s, H\Psi_s \rangle}{\langle \Psi_s, \Psi_s \rangle} ,$$

(29.5)
$$\beta = \frac{\langle \Psi_a, H\Psi_a \rangle}{\langle \Psi_a, \Psi_a \rangle} .$$

We have supposed that $\psi_1$ and $\psi_2$ are different solutions of the same eigenvalue problem. It follows that they are orthogonal:

$$\langle \psi_1, \psi_2 \rangle = 0 .$$

We have also assumed $\psi_1$ and $\psi_2$ to be normed:

$$\langle \psi_1, \psi_1 \rangle = 1 ; \quad \langle \psi_2, \psi_2 \rangle = 1 .$$

Now the scalar products occuring in (29.4) and (29.5) can easily be calculated:

$$\langle \Psi_s, \Psi_s \rangle = \langle \Psi + \Psi', \Psi + \Psi' \rangle$$
$$= \langle \Psi, \Psi \rangle + \langle \Psi, \Psi' \rangle + \langle \Psi', \Psi \rangle + \langle \Psi', \Psi' \rangle .$$

In this sum both middle terms are zero, and the outer terms $\langle \Psi, \Psi \rangle$ and $\langle \Psi', \Psi' \rangle$ are both equal to 1. Hence

$$\langle \Psi_s, \Psi_s \rangle = 2 .$$

Just so, one obtains

$$\langle \Psi_s, H \Psi_s \rangle = \langle \Psi + \Psi', H \Psi + H \Psi' \rangle$$
$$= \langle \Psi, H \Psi \rangle + \langle \Psi, H \Psi' \rangle$$
$$+ \langle \Psi', H \Psi \rangle + \langle \Psi', H \Psi' \rangle .$$

In this sum, the first and last term are equal: the scalar product remains unchanged if the two electrons are interchanged. Just so, the two middle terms are equal. Hence,

$$\langle \Psi_s, H \Psi_s \rangle = 2 \langle \Psi, H \Psi \rangle + 2 \langle \Psi, H \Psi' \rangle .$$

Now (29.4) becomes

(29.6) $$\alpha = \langle \Psi, H \Psi \rangle + \langle \Psi, H \Psi' \rangle .$$

Just so, (29.5) yields

(29.7) $$\beta = \langle \Psi, H \Psi \rangle - \langle \Psi, H \Psi' \rangle .$$

The quantities $\alpha$ and $\beta$ are the corrected energy values. Their mean is just the mean value of the energy in the state $\Psi$:

$$\frac{\alpha + \beta}{2} = \langle \Psi, H \Psi \rangle .$$

On the other hand, the difference between the two energy values is $2w$, where

(29.8) $$w = \langle \Psi, H \Psi' \rangle$$
$$= \langle \Psi, H_0 \Psi' \rangle + \langle \Psi, W \Psi' \rangle .$$

The first term is easy:

$$\langle \Psi, H_0 \Psi' \rangle = \langle \Psi, (E_1 + E_2) \Psi' \rangle$$
$$= (E_1 + E_2) \langle \psi_1(q_1) \psi_2(q_2), \psi_1(q_2) \psi_2(q_1) \rangle$$
$$= (E_1 + E_2) \langle \psi_1(q_1), \psi_2(q_1) \rangle \cdot \langle \psi_2(q_2), \psi_1(q_2) \rangle$$
$$= 0 ,$$

because $\psi_1$ and $\psi_2$ are orthogonal. Hence (29.8) reduces to

(29.9)
$$w = \langle \Psi, W \Psi' \rangle$$
$$= \int \psi_1^*(q_1) \psi_2^*(q_2) \frac{e^2}{r_{12}} \psi_1(q_2) \psi_2(q_1) \, dq_1 \, dq_2 .$$

This integral is called the "Exchange Integral".

The factor $1/r_{12}$ is large if $q_2$ is nearly equal to $q_1$. In this case the integrand is positive, because the product

$$\psi_1^*(q_1) \psi_1(q_1) \psi_2^*(q_1) \psi_2(q_1)$$

is never negative. Hence the exchange integral $w$ will be positive in most cases. This means: As a rule the symmetric state $\Psi_s$ will have a higher energy than the antisymmetric state $\Psi_a$.

Heisenberg's calculation of the integrals in (29.6) and (29.7) for the Helium atom yielded the order of magnitude of the symmetric and antisymmetric terms in good accordance with experience[2]. The ground state of Helium, which is a symmetric state, was calculated by Hylleraas[3] and later on, with greater accuracy, by Bazley[4], in perfect accordance with spectroscopic measurements.

The states $\Psi_s$ and $\Psi_a$ may be distinguished by a *Symmetry Character* $\chi$, which has the value $+1$ for symmetric states $\Psi_s$ and $-1$ for antisymmetric states $\Psi_a$. It is easy to obtain a *selection rule* for the symmetry character $\chi$. If we multiply a symmetric or antisymmetric eigenfunction $\Psi$ by $\Sigma x$ or $\Sigma y$ or $\Sigma z$, the result will be a function of the same kind. In the expansion of such a product only functions having the same symmetry character $\chi$ can occur. Hence the selection rule for $\chi$ reads

$$\chi \rightarrow \chi .$$

[2] W. Heisenberg: Über die Spektra von Atomsystemen mit zwei Elektronen. Zeitschr. f. Physik **39**, p. 499 (1926).
[3] E. A. Hylleraas: Zeitschrift f. Physik **54**, p. 347 (1929).
[4] N. W. Bazley: Lower bounds for eigenvalues with application to the Helium atom. Proceedings Nat. Acad. of Sciences **45**, p. 850 (1959).

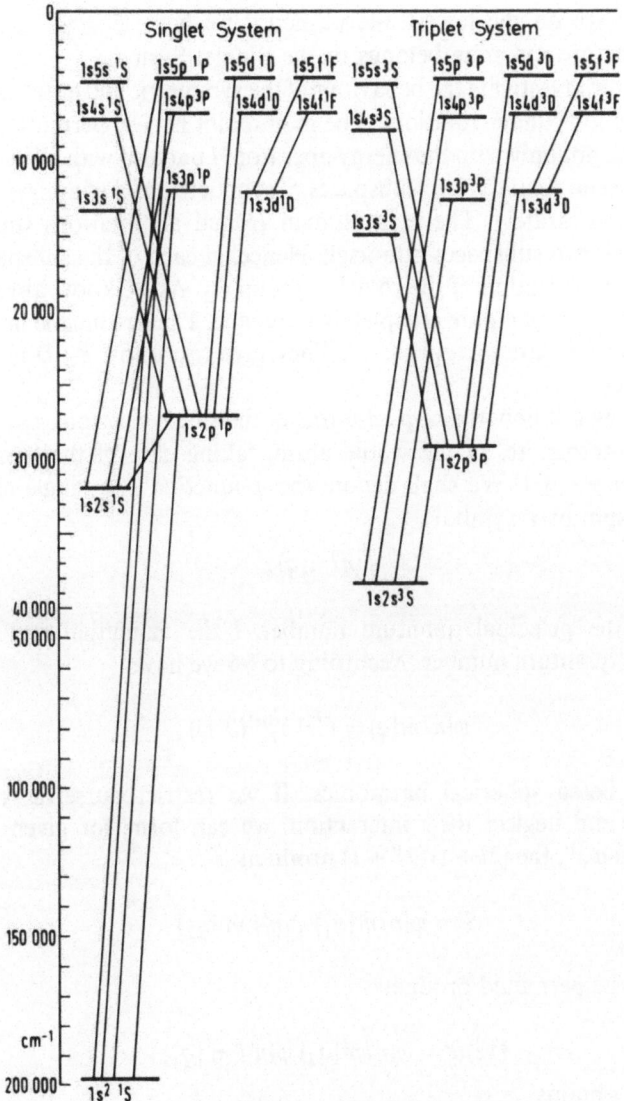

Fig. 9. Spectrum of the Helium atom

According to the theory of § 28 one would expect each of the terms to split into a triplet and a singlet. In reality the symmetric terms show only singlets, the antisymmetric terms only triplets. The reason for this phenomenon is Pauli's Exclusion Principle, which will be discussed in the next section. Thus, the singlet terms have $\chi = -1$, and combine only with singlet terms, whereas the triplet terms have $\chi = +1$ and combine only with triplet terms (see Fig. 9). In the ground state of Helium both electrons are on the lowest $s$-level, hence we have $\psi_1 = \psi_2$ and $\chi = 1$. Hence the ground state belongs to the Singlet System.

We now investigate the behaviour of the symmetric and antisymmetric eigenfunctions under rotations. The main point is: The permutation (12) commutes not only with the energy operator $H$ but also with all rotations. Hence we can treat the two subspaces of symmetric and of antisymmetric functions separately. The Hamiltonian $H$ and all rotations transform each of the two subspaces into itself. Hence, in each of the subspaces we have a representation of the rotation group $\mathcal{O}_3$. As we know already, all representations of $\mathcal{O}_3$ are completely reducible. The irreducible univalent representations are $\varrho_0, \varrho_1, \varrho_2, \ldots$. They give rise to $S$-, $P$-, $D$-terms, ... respectively.

We now ask how the $s$-, $p$-, $d$-terms of the single electrons give rise to $S$-, $P$-, $D$-terms etc. of the whole atom, taking care of the Symmetry Character $\chi = \pm 1$. We shall denote the $\psi$-function of a single electron without spin by a symbol

$$\psi(n\,l\,m\,|\,q),$$

$n$ being the principal quantum number, $l$ the azimuthal and $m$ the magnetic quantum number. According to §6 we have

$$\psi(n\,l\,m\,|\,q) = f(r)\, Y_l^{(m)}(\vartheta, \varphi),$$

the $Y_l^{(m)}$ being spherical harmonics. If we restrict ourselves to two electrons and neglect their interaction, we can form, for given values of $n$, $l$ and $n'$, $l'$, the $(2l + 1)(2l' + 1)$ products

$$\Psi = \psi(n\,l\,m\,|\,q_1)\,\psi(n'\,l'\,m'\,|\,q_2)$$

and also the permuted products

$$(12)\,\Psi = \psi(n\,l\,m\,|\,q_2)\,\psi(n'\,l'\,m'\,|\,q_1).$$

Thus one obtains

$$2(2l + 1)(2l' + 1)$$

approximate eigenfunctions, all belonging to the same energy value $E = E_1 + E_2$. If the pair $(n, l)$ is different from $(n', l')$, all these eigenfunctions are linearly independent. The linear space generated by these $2(2l + 1)(2l' + 1)$ products is a direct sum of two subspaces, one generated by the $(2l + 1)(2l' + 1)$ symmetric functions

$$\Psi_s = \Psi + (12)\,\Psi$$

and the other generated by the $(2l+1)(2l'+1)$ antisymmetric functions

$$\Psi_a = \Psi - (12)\Psi .$$

In each of the two subspaces, we have a product representation of the symmetric group, viz.

(29.10) $$\varrho_l \times \varrho_{l'} = \varrho_{l+l'} + \varrho_{l+l'-1} + \cdots + \varrho_{|l-l'|} = \Sigma \varrho_L .$$

This means: We have in the symmetric as well as in the antisymmetric subspace several sub-subspaces transforming according to the irreducible representations $\varrho_L$, where $L$ goes from $l+l'$ to $|l-l'|$. Each value of $L$ occurs twice: once in the symmetric and once in the antisymmetric subspace.

If the interaction between the two electrons is taken into account, all these sub-subspaces may belong to different energy-values. As a rule, the energy-values belonging to symmetric functions $\Psi_s$ are higher than those belonging to antisymmetric functions $\Psi_a$.

The situation is slightly more complicated if the two electrons are "in the same orbit", i.e. if $n = n'$ and $l = l'$. In this case the functions

$$(12)\,\Psi = \psi(nlm|q_2)\,\psi(nlm'|q_1)$$

can be obtained from

$$\Psi = \psi(nlm|q_1)\,\psi(nlm'|q_2)$$

by just interchanging $m$ and $m'$. This means: the functions $(12)\Psi$ are already contained among the functions $\Psi$, and we have only $(2l+1)^2$ linearly independent eigenfunctions instead of $2(2l+1)^2$. The basic symmetric eigenfunctions $\Psi_s$ are, if the indices $n$ and $l$ are dropped,

$$\Psi + (12)\,\Psi = \psi(m|q_1)\,\psi(m'|q_2) + \psi(m'|q_1)\,\psi(m|q_2)$$

and the basic antisymmetric functions $\Psi_a$ are

$$\Psi - (12)\,\Psi = \psi(m|q_1)\,\psi(m'|q_2) - \psi(m'|q_1)\,\psi(m|q_2) .$$

In the symmetric case we may assume $m \geq m'$, in the antisymmetric case $m > m'$. The eigenvalue of the operator $L_z$ is in both cases

$$M = m + m' .$$

We can now investigate how often the single possible values of $M$ occur. Let us consider the case $l = 2$. The possible values of $M$ are:

in the symmetric case

$$
\begin{aligned}
(m = \phantom{-}2) \quad M &= 4, 3, 2, 1, 0 \\
(m = \phantom{-}1) \quad M &= \phantom{4, 3, }2, 1, 0, -1 \\
(m = \phantom{-}0) \quad M &= \phantom{4, 3, 2, 1, }0, -1, -2 \\
(m = -1) \quad M &= \phantom{4, 3, 2, 1, 0, -1, }-2, -3 \\
(m = -2) \quad M &= \phantom{4, 3, 2, 1, 0, -1, -2, -3, }-4 \, ;
\end{aligned}
$$

and in the antisymmetric case

$$
\begin{aligned}
(m = \phantom{-}2) \quad M &= 3, 2, 1, 0 \\
(m = \phantom{-}1) \quad M &= \phantom{3, 2, }1, 0, -1 \\
(m = \phantom{-}0) \quad M &= \phantom{3, 2, 1, }-1, -2 \\
(m = -1) \quad M &= \phantom{3, 2, 1, 0, -1, }-3 \, .
\end{aligned}
$$

In both cases, the two largest values of $M$ (viz. 4 and 3 in the symmetric, 3 and 2 in the antisymmetric case) occur once each. The next two values (2 and 1, or 1 and 0) occur twice, and $M = 0$ occurs three times in the symmetric case. The negative values of $M$ occur just as often as the positive ones.

In order to find the irreducible representations $\varrho_L$ of the rotation group $\mathcal{O}_3$ contained in the total representation of degree $(2l + 1)^2$, we have to form sequences of $M$-values from $L$ to $-L$. Every such sequence

$$
L, L - 1, \ldots, -L
$$

gives rise to an irreducible representation $\varrho_L$. In the subspace of symmetric functions $\Psi_s$ one thus finds the representations

$$
\varrho_{2l} + \varrho_{2l-2} + \cdots + \varrho_0 \quad \text{(in our case } \varrho_4 + \varrho_2 + \varrho_0)
$$

and in the subspace of the $\Psi_a$:

$$
\varrho_{2l-1} + \varrho_{2l-3} + \cdots + \varrho_1 \quad \text{(in our case } \varrho_3 + \varrho_1).
$$

In the case of more than two electrons the situation is even more complicated. Besides the symmetric and the antisymmetric representations of the permutation group $\mathscr{S}_f$ other representations may occur. The only representations of order 1 are the symmetric and the antisymmetric representation: all others have higher order. However, there is no point in investigating all this before we have learnt about a law which considerably restricts the possibilities: Pauli's Exclusion Principle.

## § 30. The Exclusion Principle and the Periodical System[5]

As we have seen, the possible states for an outer electron in the field of a nucleus surrounded by inner electrons, are

$$1s, 2s, 2p, 3s, 3p, 3d, \dots .$$

In the case of H and $He^+$ we have a pure Coulomb field, and the energy is solely determined by the principal quantum number $n$. However, if we pass to higher atomic numbers, where the field of the nucleus is shielded by inner electrons, the $ns$-terms lie lower than the $np$-terms, the $np$-terms lower than the $nd$-terms, etc. The reason is, speaking classically, that the $s$-orbits penetrate deeper into the inner part of the atom than the $p$-orbits, so that the shielding of the field of the nucleus is more complete for the $p$-orbits than for the $s$-orbits, and so on. The usual order of the lowest terms is

$$1s, 2s, 2p, 3s, 3p, \dots \quad \text{(see Fig. 5)}.$$

The next term after $3p$ may be either $3d$ or $4s$, depending on the shielding of the field of the nucleus by the inner electrons.

One would expect that in the ground state of an atom all electrons would be on the lowest level, i.e. in the state $1s$. This is not the case: from Lithium upwards we find a quite different behaviour, closely connected with the Periodical System of the elements.

The Periodical System begins thus:

| 0 | I | II | III | IV | V | VI | VII | VIII | | |
|---|---|----|-----|----|---|----|-----|------|---|---|
| | 1. H | | | | | | | | | |
| 2. He | 3. Li | 4. Be | 5. B | 6. C | 7. N | 8. O | 9. F | | | |
| 10. Ne | 11. Na | 12. Mg | 13. Al | 14. Si | 15. P | 16. S | 17. Cl | | | |
| 18. Ar | 19. K | 20. Ca | 21. Sc | 22. Ti | 23. V | 24. Cr | 25. Mn | 26. Fe | 27. Co | 28. Ni |

The ground state of H is the $1s$-state. In the ground state of He both electrons are in the $1s$-state and the total spin is zero.

[5] For the history of the Exclusion Principle see my contribution: Exclusion Principle and Spin in the Pauli Memorial Volume (edited by Fierz and Weisskopf), p. 199. New York: Interscience Publ. 1960.

Li has two inner electrons in the $1s$-state, but the third electron, the outer or *valence electron*, is in the $2s$-state (see Fig. 5). In the case of Be there are two valence electrons, both in the $2s$-state.

If we pass to the next element B, we see that the ground state is a $P$-state. Hence, assuming that the first two electrons are in the $1s$-state and the next two in the $2s$-state, we are bound to conclude that the fifth electron is in the lowest possible $p$-state, i.e. in the $2p$-state. For C, N, O, F, Ne one also finds that the first two electrons are in the $1s$-state, the next two in the $2s$-state and all others in the $2p$-state.

The element Ne has $2+2+6$ electrons: 2 in the $1s$-orbit, 2 in the $2s$-orbit and 6 in the $2p$-orbit. The ground state of Ne is an $S$-state, i.e. it is spherically symmetric. Moreover, it belongs to the singlet system, which means that the total spin is zero.

The next element Na has again (like Li) a Hydrogen-like spectrum. The lowest term is an $S$-term, which means that the valence electron is in an $s$-state. The level of this term is higher than that of the ground state of Li, which indicates that the valence electron is not in a $2s$-state, but in a $3s$-state. The next element Mg has two $3s$-electrons, and now the same phenomena we have found in the first period (from He to F) re-occur in the second period (from Ne to Cl) on a higher energy level. With Argon the system of two $3s$-electrons and six $3p$-electrons is completed.

The next element K has again a hydrogen-like spectrum. The valence electron is in an $s$-state, and it is reasonable to assume that it is a $4s$-state. The next element Ca has two $4s$-electrons. From now on things become more complicated, because the $3d$-electrons enter the picture.

These conclusions are confirmed by Röntgen-Spectroscopy. All experimental results are in accordance with *Stoner's rule*[6]:

*In every s-orbit (with a given principal quantum number n) not more than 2 electrons can be placed, in every p-orbit not more than 6. Quite generally for every orbit with fixed quantum numbers n and l the maximum number of electrons is $2(2l+1)$.*

Accordingly the $1s$-orbit is fully occupied by 2 electrons already in the He-atom. In the Be-atom the $1s$- and $2s$-orbits are occupied, in Ne the $1s$-, $2s$-, and $2p$-orbits with $2+2+6$ electrons. In the Mg-atom the $3s$-orbit is fully occupied, in Ar the $3p$-orbit too. In Ca the $4s$-orbit is complete with 2 electrons.

From Sc on the $3d$-electrons enter the competition. In the $3d$-orbit 10 electrons can be placed, in the $4p$-orbit 6 electrons. With Xe the first "long period" of the periodic system ends: this element has

$$2+2+6+2+6+2+10+6=36$$

---

[6] E. C. Stoner: The Distribution of Electrons among Atomic Levels. Phil. Mag. **48**, p. 719 (1925).

electrons. Next comes the second "long period", which is quite similar to the first. After this, the $4f$-electrons enter the picture, and the periodicity of the system is interrupted by a group of rare metals.

In order to explain Stoner's rule, Pauli proclaimed his famous *Exclusion Principle*, which says: *No two electrons are allowed to be in the same state, including spin.* This means: Never shall two electrons have the same set of 4 quantum numbers $(n, l, j, m)$. For given values of $n$ and $l$ the quantum number $j$ can assume the two values $l \pm \frac{1}{2}$, and $m$ can have the $2j + 1$ values $j, j - 1, \ldots, -j$. This gives us all in all, for given $n$ and $l$,

$$(2l + 2) + (2l) = 2(2l + 1)$$

possible values of $j$ and $m$, as required by Stoner's rule.

In this form the Exclusion Principle is not yet invariant with respect to rotations. An invariant principle can be formulated as follows:

*The eigenfunctions of a system of electrons are required to be anti-symmetric functions of the electrons (including the spin coordinates).*

Consider e.g. the case of two $s$-electrons having the same principal quantum number $n$. Apart from the spin, the eigenfunctions of both electrons are functions of $r$, not depending on $\vartheta$ and $\varphi$. If the spin is taken into account, we have for each electron two eigenfunctions

$$\psi^{(1)} = \varphi(r)u_1 \quad \text{and} \quad \psi^{(2)} = \varphi(r)u_2 ,$$

one having its spin upwards and the other downwards. Now let $q_1$ and $q_2$ be the coordinates of the first and second electron, including spin coordinates. Instead of $q_1$ and $q_2$ we may just write 1 and 2. From the four products $\psi^{(\alpha)}(1)\,\psi^{(\beta)}(2)$ we can form just one antisymmetric linear combination, viz.

$$\Psi = \psi^{(1)}(1)\,\psi^{(2)}(2) - \psi^{(2)}(1)\,\psi^{(1)}(2) .$$

In the general case of $f$ electrons, if one starts with the eigenfunctions $\psi_1, \ldots, \psi_f$ of the single electrons, one has to form the alternating sum

(30.1) $$\Psi = \sum_P \text{sign}(P) \cdot P\{\psi_1(1)\,\psi_2(2) \ldots \psi_f(f)\} .$$

The summation extends over all permutations $P$, and $\text{sign}(P)$ is $+1$ for even, $-1$ for odd permutations. This sum is the only possible antisymmetric linear combination of the products from which it is formed.

The expression (30.1) is zero if two of the functions $\psi_k$ are equal. Hence, the exclusion principle in its original form is a consequence of the anti-symmetry.

It is reasonable to assume that the anti-symmetry not only holds for the eigenfunctions of the energy-operator $H$, but for all $\Psi$-functions at any time. The time-dependent Schrödinger equation

$$\frac{h}{i} \frac{\partial \Psi}{\partial t} + H\Psi = 0$$

implies that if a function $\Psi$ is antisymmetric at the time $t=0$ it remains so at all times.

Instead of the quantum numbers $(n, l, j, m)$ with $j = l \pm \frac{1}{2}$ and $m = j, j-1, \ldots, -j$ one can also use the quantum numbers $(n, l, m_l, m_s)$ to characterize the possible states of an electron. The number $m_l$ takes the values $l, l-1, \ldots -l$, and $m_s$ is $\pm\frac{1}{2}$. This gives us, for every pair $(n, l)$, just $2(2l+1)$ linearly independent eigenfunctions

$$(n, l, m_l, m_s) = \varphi(n, l, m_l)u_\alpha$$

(30.2)

$$(m_s = \tfrac{1}{2} \text{ for } u_1, \quad m_s = -\tfrac{1}{2} \text{ for } u_2).$$

If we have $f$ electrons, we may choose any set of $f$ linearly independent $\psi$-functions, e.g. products of the form (30.2), and from them we may form antisymmetric expressions like (30.1).

We now consider the case of a maximal number $f = 2(2l+1)$ electrons having the same quantum numbers $n, l$. From the $f$ eigenfunctions $\psi_1, \ldots, \psi_f$ we can form one single antisymmetric expression (30.1). Now if a spacial rotation $R$ is applied to the space coordinates $q_1, \ldots, q_f$ of the electrons, the expression (30.1) is transformed into another antisymmetric linear combination of the same products $\psi_1 \psi_2 \ldots \psi_f$ and hence into a multiple of itself. Hence the function $\Psi$ and its transforms $R\Psi$ all belong to a one-dimensional rotation-invariant linear space. The only one-dimensional representation of the rotation group $\mathcal{O}_3$ is the representation $\varrho_0$, hence we have $L = 0$, and the state $\Psi$ is an $S$-state.

The same reasoning holds if the rotation $R$ is applied to the spin coordinates $\sigma_1, \ldots, \sigma_f$. Hence the total spin quantum number $S$ is zero. From $L = 0$ and $S = 0$ it follows that $J = 0$. Hence the result: *A shell that is fully occupied by $2(2l+1)$ electrons is spherically symmetric and has no spin.*

As we have seen, the atoms He, Be, Ne, Mg, Ar, Ca in their ground state consist of such fully occupied shells, and the same holds for the ions $Li^+$, $Na^+$ and $K^+$. Hence the ground state of these atoms and ions is always a singlet $S$-state.

These complete shells have a particularly "closed" character if all electrons are in orbits of low energy. This is the case for the elements

He (two $1s$-electrons),

Ne (two $1s$-, two $2s$- and six $2p$-electrons),

Ar (the same and two $3s$- and six $3p$-electrons).

This explains why these elements do not form chemical compounds.

In the case of Be (two $1s$- and two $2s$-electrons) the first two shells are fully occupied, but still the element Beryllium forms chemical compounds, because the two $2s$-electrons are not as strongly bound as e.g. the two $1s$-electrons in the case of He. However, if the next shell, consisting of six $2p$-electrons is completed, all electrons are firmly bound, therefore Ne cannot form chemical compounds.

Following the inert gases He, Ne, Ar,... we find in the periodic system the alkali metals Li, Na, K, .... . Each of them has just one "valence electron" which is more loosely bound than the others and which can easily be split off. This explains why the one-valued ions $Li^+$, $Na^+$, $K^+$, $Rb^+$ and $Cs^+$ are easily formed. The inner electrons of Li, Na, K, etc. form spherically symmetric closed shells with $L = 0$, $S = 0$, and $J = 0$. A nucleus surrounded by such closed shells behaves more or less like a proton; therefore the spectra of the alkali metals are hydrogen-like (see Fig. 5).

The elements Be, Mg, Ca, etc. following the alkali metals have two valence electrons each. Their spectra have a singlet- and a triplet-system just like the spectrum of Helium (Fig. 9).

Thus, it is seen that Pauli's Exclusion Principle explains the typical chemical and spectroscopical properties of the elements in the first 3 columns of the Periodic System. In the next section we shall explain some features of more complicated spectra.


## § 31. The Eigenfunctions of the Atom

Let us first consider an atom or ion with 2 electrons only (He or $Li^+$). According to the Exclusion Principle, the eigenfunctions must change sign if the spatial *and* spin coordinates of the two electrons are interchanged. If only the space coordinates are interchanged, the eigenfunctions may be symmetric, but then they must be antisymmetric in the spin coordinates. On the other hand, if they are antisymmetric in the space coordinates, they must be symmetric in the spin coordinates.

In a good approximation, the interaction between the orbital motion and the spin may be neglected. Let

$$\psi(q_1, q_2)$$

be a symmetric or antisymmetric eigenfunction of the space coordinates $q_1$ and $q_2$ of the two electrons. The function $\psi(q_1, q_2)$ may be multiplied by any spin function of the form

(31.1)                    $a u_1 v_1 + b u_1 v_2 + c u_2 v_1 + d u_2 v_2$

where $u_1$, $u_2$ are the basis vectors in the vector space of the first electron, and $v_1$, $v_2$ of the second electron. If this expression is to be antisymmetric, the only possibility is

$$b(u_1 v_2 - u_2 v_1).$$

This expression defines a one-dimensional subspace of the four-dimensional space of all expressions (31.1). All rotations leave this subspace invariant, hence it is transformed according to the representation $\varrho_0$, i.e. the total spin $S$ is zero.

On the other hand, if the expression (31.1) is to be symmetric, it must have the form

(31.2)                    $a u_1 v_1 + b(u_1 v_2 + u_2 v_1) + d u_2 v_2$

These expressions form a 3-dimensional subspace, also invariant under rotations. The basis vectors

$$u_1 v_1, \; u_1 v_2 + u_2 v_1, \; u_2 v_2$$

are eigenfunctions of the operator $L_z$; the eigenvalues are

$$m_s = 1 \quad 0 \quad -1,$$

hence we have the representation $\varrho_1$, and the total spin is $S = 1$. Hence we obtain the final result:

*If the eigenfunction $\psi(q_1, q_2)$ without spin is symmetric, the spin $S$ is zero and the term belongs to the singlet system. If $\psi$ is antisymmetric, the spin $S$ is 1, and the term belongs to the triplet system.*

This explains the fact, already mentioned in § 29, that in the spectrum of Helium the symmetric eigenfunctions belong to singlet terms, the antisymmetric ones to multiplet terms. The same thing holds for atoms such as Be, Mg, Ca in column II of the periodic system, having just 2 valence electrons. In the ground state of such an atom the two valence electrons are in the lowest admissible $s$-state and their spins are opposite, so that the total spin is zero. The orbital momentum $L$ is also zero (because the two electrons are in $s$-states), hence the ground state is a singlet $S$-state.

In the case of more than two valence electrons the situation is more complicated. The main question is: How are the eigenfunctions transformed if the space coordinates or the spin coordinates are subjected to a rotation or to a permutation?

Let us first neglect the orbit-spin interaction and write the eigenfunctions as products

$$(31.3) \qquad \varphi_\alpha(q_1, ..., q_f) u_\lambda v_\mu ... w_\nu$$

or as linear combinations of such products, the $\varphi_\alpha$ being eigenfunctions of the space coordinates of the atoms without spin. On every energy level we have 4 commuting groups of linear transformations operating on the functions (31.3), viz.

the rotations of $q$-space,

the rotations of the Spin-Space,

the permutations of $q_1, ..., q_f$,

the permutations of $u, v, ..., w$.

We have to investigate how these four groups operate on the functions (31.3), taking care of the Exclusion Principle. There are two methods for this investigation.

The *first method* is: First consider the space functions $\varphi_\alpha(q_1, ..., q_f)$ only, on which two groups operate: the rotations and the permutations. Because the two groups commute, we can order the eigenfunctions in rectangles as in § 15. The rows of each rectangle are transformed according to an irreducible representation $\varrho_L$ of the rotation group $\mathcal{O}_3$, and the columns according to an irreducible representation $\varrho_S$ of the permutation group $\mathcal{S}_f$.

Next apply the same consideration to the $2^f$ spin functions $u_\lambda v_\mu ... w_\nu$. Once more, one can construct rectangles such that the rows of each rectangle are transformed by rotations according to an irreducible representation $\varrho_S$ of the rotation group $\mathcal{O}_3$, and the columns according to an irreducible representation $\varDelta'$ of the permutation group $\mathcal{S}_f$. As an example, consider the case of 2 electrons ($f = 2$). In this case there are just 2 rectangles:

$$\boxed{u_1 v_1, \ u_1 v_2 + u_2 v_1, \ u_2 v_2}$$

and

$$\boxed{u_1 v_2 - u_2 v_1} \ .$$

The only row of such a rectangle is transformed according to the representation $\varrho_1$ or $\varrho_0$ of the rotation group. If the two electrons are interchanged, every column of the first rectangle remains invariant, while the only column of the second rectangle changes sign.

The next step would be, to combine the space functions $\psi(q)$ obtained in the first step with the spin functions obtained in the second step, and to select those functions of the space and spin coordinates that satisfy the Exclusion Principle. The selected functions have to be antisymmetric under permutations of the space *and* spin coordinates. Now if we multiply the space functions in a column of a rectangle, which are transformed according to a representation $\Delta$ of the group $\mathscr{S}_f$, with the spin functions in a column of another rectangle, which are transformed according to $\Delta'$, we obtain a set of products which is transformed according to the product representation $\Delta \times \Delta'$. Hence the question is: does this product representation contain the antisymmetric representation $A$ as an irreducible component? According to a theorem proved in § 14 this question is equivalent to the other question: does the product

$$\Delta \times \Delta' \times A$$

contain the identical representation? This in turn is equivalent to the question: does $\Delta \times A$ contain as an irreducible component the representation $\widetilde{\Delta'}$ dual to $\Delta'$?

Now since $\Delta$ is irreducible and $A$ of degree 1 the representation $\Delta \times A$ is irreducible. Hence $\Delta \times A$ contains $\widetilde{\Delta'}$ as an irreducible component only if $\Delta \times A$ is equivalent to $\widetilde{\Delta'}$, or dual to $\Delta'$. If this is the case, one can form from any column of the first rectangle with column number $m_L$ and any column of the second rectangle with column number $m_S$ an anti-symmetric function

$$(31.4) \qquad\qquad \psi^{(m_L, m_S)}$$

of the space- and spin-coordinates. Here $m_L$ ranges from $-L$ to $+L$, and $m_S$ from $-S$ to $+S$. The set of functions (31.4) is transformed under rotations according to the representation

$$\varrho_L \times \varrho_S = \Sigma \varrho_J \quad (J = L+S, L+S-1, ..., |L-S|),$$

as in § 21.

This "first method" was applied in the earliest papers in which group representations were applied to the classification of atomic spectra, notably by von Neumann and Wigner[7].

---

[7] J. von Neumann and E. Wigner: Zur Erklärung einiger Eigenschaften der Spektren aus der Quantenmechanik des Drehelektrons II and III. Zeitschr. f. Phys. **49**, p. 73 and **51**, p. 844 (1928).

For the application of this method, a thorough knowledge of the representation theory of the symmetric group was required: one had to calculate the representations $\Delta \times A$ and $\Delta'$ explicitely and to decide whether they are dual to each other. Physicists found this method very difficult: they disliked the "group pest".

However, there exists another method, which had been applied to special cases already before the "group pest" broke out. This *second method* was systematically developed by J. C. Slater[8]. It does not use the Representation Theory of the Symmetric Group. Its principle is as follows: One does not discuss permutations of the space coordinates and spin coordinates separately, but only permutations $P$ of the electrons as a whole, and one restricts oneself from the very beginning to anti-symmetric eigenfunctions

$$(31.5) \qquad \Psi = \sum_P (\text{sign } P) \cdot P\{\varphi(q_1, \ldots, q_f) u_\lambda v_\mu \ldots w_\nu\}.$$

Rotations of $q$-space as well as of spin-space transform antisymmetric eigenfunctions into antisymmetric ones. Hence one has, in the linear space generated by the functions (36.5), two commuting groups of linear transformations induced by the $q$-rotations and spin-rotations. Hence we can form, once more, rectangles of $(2L+1)(2S+1)$ functions $\psi^{(m_L, m_S)}$ such that the rows of each rectangle are transformed under $q$-rotations according to $\varrho_L$, while the columns are transformed under spin-rotations according to $\varrho_S$. If the rotations $R$ are applied to the space- and spin-coordinates simultaneously, the functions $\psi^{(m_L, m_S)}$ are transformed just like products

$$\varphi^{(m_L)} \cdot \psi^{(m_S)}$$

which means that they are transformed according to the representation

$$\varrho_L \times \varrho_S = \Sigma\, \varrho_J = \varrho_{L+S} + \varrho_{L+S-1} + \cdots + \varrho_{|L-S|}.$$

The single terms with $J = L+S, \ldots, |L-S|$ coincide as long as the interaction between orbital motion and spin is neglected. If the interaction is taken into account, the $2L+1$ or $2S+1$ terms of the multiplet may become separate.

From the preceding investigation we see that the Exclusion Principle never excludes a part of a multiplet: *The whole multiplet is either excluded or admitted.*

We shall now use Slater's method to answer the following question: What kinds of multiplets can result from electrons in given orbits?

---

[8] J. C. Slater: Phys. Rev. **34** (2), p. 1293 (1929).

To begin with, let us neglect the interaction between the electrons. The eigenfunctions can now be written as products:

$$\Psi_\alpha = \psi(k_1|q_1)\,\psi(k_2|q_2)\ldots\psi(k_f|q_f)\,.$$

Here the symbols $k$ stand for triples of quantum numbers $(n, l, m_l)$. Hence the function $\Psi_\alpha$ is determined by a list of the triples

$$k = (n, l, m_l)$$

occuring in the product $\Psi_\alpha$. In order to obtain a term of the sum (31.5), we have to multiply $\Psi_\alpha$ by a product $u_\lambda v_\mu \ldots w_\nu$, in which each of the indices $\lambda, \mu, \ldots$ can be either 1 or 2. Instead of the numbers 1 and 2 we can also use the symbols $+$ and $-$, namely $+$ for $m_s = +\frac{1}{2}$ and $-$ for $m_s = -\frac{1}{2}$. These symbols may be combined with the triples $(n, l, m_l)$ so as to obtain quadruples

$$(n, l, m_l, +) \quad \text{or} \quad (n, l, m_l, -)\,.$$

Thus, every antisymmetric eigenfunction (30.5) will be determined by $f$ quadruples $(n, l, m_l, \pm)$. For instance, the eigenfunction

$$(31.6) \qquad \Psi = \sum_P (\operatorname{sign} P) \cdot P\{\psi(211|q_1)\,\psi(210|q_2)u_1 v_2\}$$

will be represented by the symbol

$$(31.7) \qquad\qquad\qquad (211+)(210-)\,.$$

Since the function $\Psi$ may be multiplied by $-1$, the order of the factors of the product does not matter: the symbol

$$(210-)(211+)$$

represents the same state of the atom. Two equal factors $(n, l, m_l, m_s)$ cannot occur, because otherwise the sum (31.6) would become zero.

Now if the pairs $(n, l)$ determining the energy levels of the single electrons without spin are given, and if each $m_l$ takes on the values $l, l-1, \ldots, -l$ and each $m_s$ the values $+$ or $-$, we obtain a certain number of symbols like (31.7), each determining a function $\Psi$. For each of these symbols we can calculate the sums

$$M_L = \Sigma\, m_l \quad \text{and} \quad M_S = \Sigma\, m_s$$

and then make a list of the pairs of values $(M_L, M_S)$. For instance in the case of three $2p$-electrons the first half of the list, containing the positive values of $M_S$, would look as follows. The second half of the list may be omitted, because it can be obtained from the first half by changing the signs of $M_L$ and $M_S$.

|  | $M_L$ | $M_S$ |
|---|---|---|
| $(2\,1\,1\,+)(2\,1\ \ \ 0\,+)(2\,1\,-1\,+)$ | 0 | $\frac{3}{2}$ |
| $(2\,1\,1\,+)(2\,1\ \ \ 0\,+)(2\,1\ \ \ 1\,-)$ | 2 | $\frac{1}{2}$ |
| $(2\,1\,1\,+)(2\,1\ \ \ 0\,+)(2\,1\ \ \ 0\,-)$ | 1 | $\frac{1}{2}$ |
| $(2\,1\,1\,+)(2\,1\ \ \ 0\,+)(2\,1\,-1\,-)$ | 0 | $\frac{1}{2}$ |
| $(2\,1\,1\,+)(2\,1\,-1\,+)(2\,1\ \ \ 1\,-)$ | 1 | $\frac{1}{2}$ |
| $(2\,1\,1\,+)(2\,1\,-1\,+)(2\,1\ \ \ 0\,-)$ | 0 | $\frac{1}{2}$ |
| $(2\,1\,1\,+)(2\,1\,-1\,+)(2\,1\,-1\,-)$ | $-1$ | $\frac{1}{2}$ |
| $(2\,1\,0\,+)(2\,1\,-1\,+)(2\,1\ \ \ 1\,-)$ | 0 | $\frac{1}{2}$ |
| $(2\,1\,0\,+)(2\,1\,-1\,+)(2\,1\ \ \ 0\,-)$ | $-1$ | $\frac{1}{2}$ |
| $(2\,1\,0\,+)(2\,1\,-1\,+)(2\,1\,-1\,-)$ | $-2$ | $\frac{1}{2}$ |

From the pairs $(M_L, M_S)$ we now form rectangles as large as possible, $M_L$ ranging from $+L$ to $-L$ and $M_S$ from $+S$ to $-S$ in every rectangle. We start with the largest value $S$ of $M_S$ (in our case $S = \frac{3}{2}$) and look up the largest value $L$ of $M_L$ corresponding to this value of $M_S$ (in our case it is $L = 0$). Starting with this largest value $L$, we always find a whole sequence of values of $M_L$:

$$L, L-1, \ldots, -L$$

corresponding to the same value of $M_S$. In our case the sequence would consist of $M_L = 0$ only. Thus we obtain our first rectangle

$$
\begin{array}{|c|}
\hline
0, \quad \frac{3}{2} \\
0, \quad \frac{1}{2} \\
0, -\frac{1}{2} \\
0, -\frac{3}{2} \\
\hline
\end{array}
$$

giving $L = 0$ and $S = \frac{3}{2}$.

We now strike out these pairs. Among the remaining pairs the largest value of $M_S$ is $S = \frac{1}{2}$. The largest value of $M_L$ corresponding to $M_S = \frac{1}{2}$

is $L = 2$. This gives us another rectangle ($L = 2, S = \frac{1}{2}$)

$$(2, \ \tfrac{1}{2}) \ (1, \ \tfrac{1}{2}) \ (0, \ \tfrac{1}{2}) \ (-1, \ \tfrac{1}{2}) \ (-2, \ \tfrac{1}{2})$$
$$(2, -\tfrac{1}{2}) \ (1, -\tfrac{1}{2}) \ (0, -\tfrac{1}{2}) \ (-1, -\tfrac{1}{2}) \ (-2, -\tfrac{1}{2})$$

Striking out these pairs, we are left with a single rectangle ($L = 1, S = \frac{1}{2}$):

$$(1, \ \tfrac{1}{2}) \ (0, \ \tfrac{1}{2}) \ (-1, \ \tfrac{1}{2})$$
$$(1, -\tfrac{1}{2}) \ (0, -\tfrac{1}{2}) \ (-1, -\tfrac{1}{2}).$$

Hence three $2p$-electrons give rise to 3 possible terms:

$$^4S \ (L = 0, S = \tfrac{3}{2}) \quad \text{(pronounce: quadruplet } S\text{)},$$

$$^2D \ (L = 2, S = \tfrac{1}{2}) \quad \text{(pronounce: doublet } D\text{)},$$

$$^2P \ (L = 1, S = \tfrac{1}{2}) \quad \text{(pronounce: doublet } P\text{)}.$$

The first term $^4S$ has an upper index $2S + 1 = 4$, because it belongs to the quadruplet system. The term is not really a quadruplet, but a singlet, because $L$ is zero. The other two terms are doublets.

Of course, the particular value of the principal quantum number $n = 2$ does not matter. Quite generally, if electrons having the same pair of quantum numbers $(n, l)$ are called *equivalent*, we may conclude:

*Three equivalent p-electrons give rise to the terms $^4S, \ ^2D, \ ^2P$.*

In just the same way all possible cases can be treated. In the case of 2 electrons the results are

1. *For two inequivalent electrons $(n, l)$ and $(n', l')$ singlet terms with $L = l + l', l + l' - 1, ..., |l - l'|$ and triplet terms with the same L-values are possible, the singlet terms $(S = 0)$ being symmetric in the space coordinates and the triplet terms $(S = 1)$ antisymmetric.*

2. *For two equivalent electrons $(n, l)$ one has only symmetric singlet terms with*

$$L = 2l, 2l - 2, ..., 0$$

*and antisymmetric triplet terms with*

$$L = 2l - 1, 2l - 3, ..., 1.$$

These results can easily be obtained by directly writing down all the eigenfunctions. Those that are symmetric in the space coordinates are antisymmetric in the spin coordinates and have the form

$$\varphi_\beta(q_1, q_2) \cdot (u_1 v_2 - u_2 v_1)$$

whereas the others are antisymmetric in the space coordinates and symmetric in the spin coordinates; they may be written as $\varphi_\beta(q_1, q_2)$ multiplied by

$$u_1 v_1, \quad u_1 v_2 + u_2 v_1, \quad u_2 v_2$$

in accordance with our earlier results.

For more than 2 electrons, the following general rules may be formulated, which are direct consequences of the method of Slater:

**Rule 1.** *A fully occupied shell $(n, l)$, in which all possible combinations $m_l = l, l-1, \ldots -l$ and $m_s = \pm \frac{1}{2}$ occur just once, can be left out of account in calculating the possible values of $M_L$ and $M_S$.*

*Proof.* For these values of $m_l$ and $m_s$ we have $\Sigma\, m_l = 0$ and $\Sigma\, m_s = 0$.

**Rule 2.** *If outside the fully occupied shells one has two (or more) inequivalent sets of electrons, one can first treat the two sets separately and calculate*

$$\Sigma\, m_l = M_L', \quad \Sigma\, m_s = M_S'$$

*for the first set, and just so*

$$\Sigma\, m_l = M_L'', \quad \Sigma\, m_s = M_S''$$

*for the second set. The pairs $(M_L', M_S')$ can be arranged in rectangles of $(2L' + 1)(2S' + 1)$ pairs each, and the pairs $M_L'', M_S''$ in rectangles of $(2L'' + 1)(2S'' + 1)$. Thus one obtains some pairs $(L', S')$ for the first set of electrons, and some pairs $(L'', S'')$ for the second set. Finally, every $L'$ is combined with every $L''$ to form a sequence of total azimuthal quantum numbers*

$$L = L' + L'', L' + L'' - 1, \ldots, |L' - L''|$$

*and every $S'$ with every $S''$ to form a sequence of spin quantum numbers*

$$S = S' + S'', S' + S'' - 1, \ldots, |S' - S''|.$$

**Example.** What terms are possible for a Nitrogen atom (7 electrons) if the $1s$- and $2s$-orbits are fully occupied and if, in addition to these closed shells, one has two $2p$-electrons and one $3s$-electron?

The two 2p-electrons give rise to three terms

$$^3P \ (L=1, S=1),$$
$$^1D \ (L=2, S=0),$$
$$^1S \ (L=0, S=0).$$

If these are combined with the 3s-electron, one obtains four terms

$$2p^2 \, 3s \, {}^4P \ (L=1, S=1\tfrac{1}{2}),$$
$$2p^2 \, 3s \, {}^2P \ (L=1, S=\tfrac{1}{2}),$$
$$2p^2 \, 3s \, {}^2D \ (L=2, S=\tfrac{1}{2}),$$
$$2p^2 \, 3s \, {}^2S \ (L=0, S=\tfrac{1}{2}).$$

According to this general method, it is easy to write down all possible terms of the whole atom as soon as the possibilities for groups of equivalent electrons are known. The maximum number of equivalent s-electrons is 2, of p-electrons 6, etc. The s-electrons offer no difficulty at all, since two of them already form a closed shell. The cases of one, two or three equivalent p-electrons have been discussed already. Next suppose we have 4 equivalent p-electrons. The labour of writing down all possible sequences of four symbols $(n, l, m_s, m_l)$ can be simplified by writing down the two missing symbols needed to obtain a closed shell. Namely: for a closed shell $\Sigma \, m_l$ and $\Sigma \, m_s$ are always zero, so the sum $\Sigma \, m_l$ for the missing two symbols is equal to the sum for the four symbols actually occurring, with the opposite sign. It follows that the possible terms $(L, S)$ for 4 equivalent p-electrons are just the same as those for 2 equivalent p-electrons.

Obviously, the same considerations hold in all cases. Hence one obtains

**Rule 3.** *Four equivalent p-electrons yield the same terms as two, and five the same as one single p-electron. Just so, six equivalent d-electrons yield the same terms as four, seven the same as three, eight the same as two, and nine the same as one single d-electron.*

**Example.** Of what kind is the ground state of the Oxygen atom?

If all eight electrons are successively brought into the lowest possible states, one obtains the symbol

$$1s^2 \, 2s^2 \, 2p^4 \, .$$

Apart from closed shells, we have four $2p$-electrons. The possible terms are the same as for two $2p$-electrons, viz.

$$^1S, \,^3P, \,^1D\,.$$

According to an empirical rule, the lowest terms are those of highest multiplicity; in our case this is the triplet $P$ term. So the ground state has the symbol

$$1s^2\, 2s^2\, 2p^4\, {}^3P\,.$$

The quantum number $w = (-1)^{\Sigma l}$ is $+1$, since $\Sigma l$ is 4 (compare § 21).

For a more thorough discussion of the terms and spectral lines of the elements I may refer the reader to the classical book of F. Hund: Linienspektren und periodisches System der Elemente, Berlin 1927.

## § 32. The Calculation of the Energy Values

Until now we have neglected the interaction between the electrons. This gives us the right kinds of terms, but for the energy values the approximation is not good. A better approximation may be obtained as follows. For every single electron the interaction with the others is replaced by a shielding of the Coulomb field of the nucleus. A good method for calculating the amount of this shielding is Hartree's "Method of the self-consistent field"[9]. For every single electron the potential of the shielded field is determined by the following condition: If one calculates the eigenfunctions $\varphi_\alpha(q)$ of any particular electron (number $\alpha$), and if one next calculates the total charge density of all electrons except one (number $\beta$), i.e. the sum

$$-e \sum_{\alpha \neq \beta} \varphi_\alpha^* \varphi_\alpha$$

then the potential of this charge together with the potential of the nucleus is required to give just the same potential field from which one started. A potential field satisfying this condition is called "self-consistent".

In most cases the calculation may be simplified by averaging the charge density over a sphere of radius $r$ around the nucleus; this proves to be a reasonably good approximation. The potential fields thus obtained are spherically symmetric.

It turns out that these "self-consistent fields" can be determined to a sufficient degree of accuracy by successive approximations. In most cases

---

[9] D. R. Hartree: Proc. Cambr. Phil. Soc. **24**, p. 89 (1928).

these energy values agree well with the observed spectral terms. Therefore one can assume that the product of Hartree's eigenfunctions

$$(32.1) \qquad \varphi(q) = \varphi_1(q_1)\, \varphi_2(q_2) \ldots \varphi_f(q_f)$$

may be used as a reasonable first approximation of the eigenfunctions of the atom.

The product (32.1) is only one out of

$$(2l+1)(2l'+1)\ldots$$

linearly independent products, all belonging in first approximation to the same energy value. Under rotations these functions $\varphi(q)$ are transformed according to a product representation

$$\varrho_l \times \varrho_{l'} \times \cdots = \Sigma\, \varrho_L.$$

In a second approximation, the irreducible subspaces $\mathscr{V}_L$ of the product space may belong to different energy values. To obtain these, we have to use Perturbation Theory. The calculation may be made as follows.

As in § 31, we first form products of the functions $\varphi(q)$ with arbitrary functions of the spin-coordinates

$$(32.2) \qquad \varphi(q_1, \ldots, q_f)\, u_\lambda v_\mu \ldots w_\nu$$
$$= \psi(nlm_l m_s | q_1, \sigma_1) \cdot \psi(n'l'm_l'm_s' | q_2, \sigma_2) \ldots .$$

In first approximation, all these products belong to the same energy value. In the linear space generated by these products we form the subspace $\mathscr{V}$ of antisymmetric functions

$$\Psi_\alpha = \sum_P (\operatorname{sign} P) \cdot P\{\varphi(q) u_\lambda v_\mu \ldots w_\mu\}$$
$$= \sum_P (\operatorname{sign} P) \cdot P\{\psi(nl \ldots | q_1, \sigma_1)\, \psi(n'l' \ldots | q_2, \sigma_2) \ldots\} .$$

Every function $\Psi_\alpha$ of this kind is determined by a combination of symbols

$$(nlm_l m_s)(n'l'm_l'm_s')\ldots ,$$

and functions belonging to different combinations are orthogonal. Hence the functions $\Psi_\alpha$ form an orthogonal basis of the subspace $\mathscr{V}$.

We now apply the energy operator $H$ (with interaction between the electrons, but without spin-orbit and spin-spin interaction) to the functions $\Psi_\beta$. The result $H\Psi_\beta$ is again antisymmetric. In the space of all antisymmetric functions we take a basis consisting first of the functions $\Psi_\alpha$ spanning the subspace $\mathscr{V}$, and of other functions $\Psi'_j$ belonging to the subspace $\mathscr{V}^\perp$ orthogonal to $\mathscr{V}$. The expansion of $H\Psi_\beta$ now reads

$$H\Psi_\beta = \Sigma\, \Psi_\alpha h_{\alpha\beta} + \cdots$$

the terms $+\cdots$ belonging to $\mathscr{V}^\perp$. According to Perturbation Theory, we have to diagonalize the matrix $(h_{\alpha\beta})$. The diagonal elements are the energy values in second approximation.

The diagonalization of the matrix $(h_{\alpha\beta})$ is considerably simplified by the fact that there are four operators commuting with $H$ and with each other, viz:

$\mathscr{L}^2$, the square of the total orbital angular momentum, having eigenvalues $L(L+1)$,

$L_z$, the orbital angular momentum in the $z$-direction, having eigenvalues

$$M_L = \Sigma\, m_l\,,$$

$\mathscr{S}^2$, the square of the spin momentum, having eigenvalues $S(S+1)$,

$S_z$, the spin momentum in the $z$-direction, having eigenvalues

$$M_S = \Sigma\, m_s\,.$$

These four operators can be diagonalized simultaneously with $H$, i.e. we can find new functions $\Phi_\alpha$ that are eigenfunctions of $H$, $\mathscr{L}^2$, $L_z$, $\mathscr{S}^2$, and $S_z$.

The easiest procedure to find such eigenfunctions $\Phi_\alpha$ without solving secular equations of high degrees is as follows. First one remarks that $L_z$ and $S_z$ are already diagonalized by the original basis functions $\Psi_\alpha$. Every $\Psi_\alpha$ is an eigenfunction of $L_z$ belonging to a definite eigenvalue $\Sigma\, m_l$, and at the same time an eigenvalue of $S_z$ belonging to a definite eigenvalue $\Sigma\, m_s$. Hence, to every pair of values $(M_L, M_S)$, one can collect those functions $\Psi_\alpha$ that belong to this pair of values.

Next, one rearranges the pairs $(M_L, M_S)$ in rectangles, as in § 32. In every rectangle, $M_L$ goes from $L$ to $-L$, and $M_S$ goes from $S$ to $-S$. To every rectangle corresponds an eigenvalue $L(L+1)$ of $\mathscr{L}^2$ and an eigenvalue $S(S+1)$ of $\mathscr{S}^2$. The eigenfunctions can be calculated, if necessary, by the methods of § 21, i.e. by the Clebsch-Gordan series, but in most cases it is not even necessary to calculate them, as we shall see presently.

In most cases, there is only one eigenfunction $\Phi$ corresponding to each given set of quantum numbers $(L, S, M_L, M_S)$. In these cases, it is not even necessary to solve a secular equation. One just calculates the mean value of the energy in any state $\Phi_\alpha$:

$$\langle \Phi_\alpha, H\Phi_\alpha \rangle = \int \Phi_\alpha^* H\Phi_\alpha \, dV = E_\alpha$$

by integration over the space coordinates and summation over the spin coordinates, and one has the energy value $E_\alpha$ in second approximation. In exceptional cases one has to solve a secular equation of degree 2.

The calculation can be even more simplified by computing traces. The trace of a matrix $A$

$$T(A) = \Sigma \, a_{kk}$$

is independent of the choice of the basis vectors. Hence if one collects all eigenfunctions $\Psi_\beta$ belonging to a definite pair of values

$$M_L = \Sigma \, m_l, \qquad M_S = \Sigma \, m_s$$

the sum of the diagonal elements

$$h_{\beta\beta} = \langle \Psi_\beta, H\Psi_\beta \rangle$$

is equal to the sum of the diagonal elements

$$\langle \Phi_\alpha, H\Phi_\alpha \rangle = E_\alpha$$

belonging to the same values of $M_L$ and $M_S$. Hence, in most cases, it is perfectly sufficient to calculate the diagonal elements $h_{\beta\beta}$ of the matrix $H$.

For further details of the calculation I may refer to the paper of J. C. Slater quoted in Footnote 8 on p. 165.

## § 33. Pure Spin Functions and their Transformation under Rotations and Permutations

In § 31 and § 32 we have used Slater's method to determine the possible terms of an atom. This method leaves two questions open, namely: How are the eigenfunctions $\varphi(q)$ transformed under permutations, if the spin is left out of account? And by what kind of spin functions are these functions $\varphi(q)$ to be multiplied if one wants to obtain antisymmetric functions $\Psi$ of all coordinates?

To answer these questions we have to use the "first method" explained in § 31, i.e. we must investigate the transformation of space functions and spin functions separately. If a set of space functions $\varphi(q)$ is transformed according to a representation $\Delta$ of the permutation group $\mathscr{S}_f$, and if they are multiplied by spin functions which are transformed according to a representation $\Delta'$, the condition we found in § 31 is that $\Delta'$ must be dual to $\Delta \times A$, where $A$ is the antisymmetric representation of $\mathscr{S}_f$. As we shall see, the dual representation to $\Delta'$ is $\Delta'$ itself, so that the necessary relation between $\Delta$ and $\Delta'$ can also be written as

$$\Delta \times A \sim \Delta' \quad (\sim \text{means: equivalent})$$

or, if we multiply both sides by $A$, as

$$(33.1) \qquad\qquad\qquad \Delta \sim \Delta' \times A.$$

Hence, as soon as $\Delta'$ is known, $\Delta$ is also known. We now proceed to determine $\Delta'$. It turns out that $\Delta'$ is completely determined by the spin quantum number $S$.

All spin functions are linear combinations of $2^f$ products

$$(33.2) \qquad\qquad u_\lambda v_\mu \dots w_\nu \quad (\lambda, \mu, \dots = 1, 2)$$

They form a 2-dimensional vector space $\mathscr{W}$, on which two groups operate: first the group of permutations of the letters $u, v, \dots, w$, secondly the group of simultaneous unitary transformations of the $u_\lambda, v_\mu, \dots, w_\nu$. Hence we have in $\mathscr{W}$ a representation $\pi$ of the permutation group and a representation $\delta$ of the unitary group $SU(2)$. The two representations commute, hence the basis vectors of the irreducible subspaces can be ordered in rectangles

$$\begin{array}{ccc} v_{11} & \dots & v_{1m} \\ \vdots & & \\ v_{k1} & \dots & v_{km} \end{array}$$

such that the rows of every rectangle are transformed according to an irreducible representation of $\mathscr{S}_f$, whereas the columns are transformed according to an irreducible representation $\varrho_S$ of $SU(2)$.

As an example, let us consider the case $f = 2$. There are only two irreducible representations of $\mathscr{S}_2$: the symmetric and the antisymmetric representation. There are three symmetric bilinear functions of $u_\lambda$ and $v_\mu$ and one antisymmetric function, viz.

$$\begin{array}{c} u_1 v_1 \\ u_1 v_2 + u_2 v_1 \\ u_2 v_2 \end{array} \quad \text{and} \quad \boxed{u_1 v_2 - u_2 v_1}.$$

Under the group SU(2), the three symmetric functions are transformed according to $\varrho_1$, and the antisymmetric one according to $\varrho_0$. Thus, to every value of the spin quantum number $S$, we have just one rectangle and just one representation $\Delta'$, viz. the symmetric one for $S = 1$ and the antisymmetric one for $S = 0$.

The same thing holds for arbitrary values of $f$. This is a consequence of the following Main Theorem:

*All matrices commuting with the matrices of the representation $\pi$ are linear combinations of the matrices of the representation $\delta$.*

*Proof.* Let a linear transformation of the vector space $\mathcal{V}$ into itself be given by

$$(33.3) \qquad T u_\lambda v_\mu \ldots w_\nu = \Sigma \, c_{\lambda'\lambda,\mu'\mu,\ldots,\nu'\nu} u_{\lambda'} v_{\mu'} \ldots w_{\nu'} .$$

Now if $T$ permutes with all permutations of the letters $u, v, \ldots, w$, this means that the coefficients $c_{\lambda'\lambda,\mu'\mu,\ldots,\nu'\nu}$ remain unchanged if the pairs of indices $\lambda'\lambda, \ldots, \nu'\nu$ are subjected to an arbitrary permutation.

To simplify the notation, let us write $l$ instead of the pair $\lambda'\lambda$, and $m$ instead of $\mu'\mu$, etc. Now $c_{l,m,\ldots,n}$ must be symmetric in all indices:

$$c_{lm\ldots n} = c_{ml\ldots n} = \text{etc} .$$

The representation $\delta$ consists of all linear transformations induced by unitary transformations

$$u'_\lambda = \Sigma \, u_{\lambda'} c_{\lambda'\lambda} ; \qquad v'_\mu = \Sigma \, v_{\mu'} c_{\mu'\mu} ; \quad \ldots$$

with

$$\begin{pmatrix} c_{11} & c_{12} \\ c_{21} & c_{22} \end{pmatrix} = \begin{pmatrix} \alpha & \beta \\ -\beta^* & \alpha^* \end{pmatrix} \quad \text{and} \quad \alpha\alpha^* + \beta\beta^* = 1 .$$

It follows that the products $u_\lambda v_\mu \ldots$ are transformed as follows

$$u'_\lambda v'_\mu \ldots w'_\nu = \Sigma \, u_{\lambda'} v_{\mu'} \ldots w_{\nu'} c_{\lambda'\lambda} c_{\mu'\mu} \ldots c_{\nu'\nu} .$$

The coefficients of these transformations are

$$c_{\lambda'\lambda,\mu'\mu,\ldots,\nu'\nu} = c_{\lambda'\lambda} c_{\mu'\mu} \ldots c_{\nu'\nu}$$

or in our shorter notation

$$(33.4) \qquad c_{lm\ldots n} = c_l c_m \ldots c_n .$$

We have to prove, that all symmetrical tensors $c_{lm...n}$ are linear combinations of the special tensors (33.4), or, what amounts to the same thing, that all linear equations

(33.5)  $$\Sigma \gamma_{lm...n} c_{lm...n} = 0$$

which hold for the special tensors (33.4) also hold for any symmetric tensor $c_{lm...n}$.

Let us put

(33.6)
$$c_{11} = \alpha = a_1 + i a_2, \quad c_{12} = \beta = a_3 + i a_4,$$
$$c_{22} = \alpha^* = a_1 - i a_2, \quad c_{21} = -\beta^* = -a_3 + i a_4.$$

Now if an equation

(33.7)  $$\Sigma \gamma_{lm...n} c_l c_m \cdots c_n = 0$$

holds for all $c_l = c_{\lambda' \lambda}$, we may substitute (33.6) into this equation, thus obtaining an equation which holds for all real numbers $a_1, a_2, a_3, a_4$ with

$$a_1^2 + a_2^2 + a_3^2 + a_4^2 = 1.$$

Because the Eq. (33.7) is homogeneous, it remains valid if all $a_k$ are multiplied by one and the same real factor $\varrho$; hence the equation holds identically in $a_1, a_2, a_3, a_4$. Now the $c_{ij}$ are linear functions of the $a_k$ and conversely, hence the Eq. (33.7) holds identically in the $c_{ij}$, i.e. all coefficients in this equation are zero. Now the total coefficient of any product $c_l c_m \cdots c_n$ in (33.7), i.e. the sum of all coefficients $\gamma_{lm...n}$ obtained from one $\gamma_{lm...n}$ by permutations of the indices $lm \ldots n$, is equal to the total coefficient of $c_{lm...n}$ in (33.5). Hence these total coefficients are zero, i.e. (33.5) holds for all symmetric tensors $c_{lm...n}$, which is just what we wanted to prove.

The Main Theorem just proved holds not only for the unitary group SU(2), but also for the special linear group SL(2) and the general linear group GL(2), no matter whether we consider real or complex coefficients. The proofs for SL(2) and GL(2) are even simpler than for SU(2), because the passage from complex $c_{ij}$ to real $a_k$ is not necessary in the cases of SL(2) and GL(2).

The Main Theorem can also be generalized to the groups SL($n$) and GL($n$). The proof is just the same as in the case $n = 2$.

The Main Theorem also holds for SU($n$), but the proof is a little more complicated. The main point is the passage from (33.7) to (33.5). We have to prove: if an Eq. (33.7) holds for all unitary matrices $(c_{\lambda' \lambda})$, it holds for all matrices $(c_{\lambda' \lambda})$, i.e. it holds identically in the $c_l = c_{\lambda' \lambda}$. The proof can be given by using the infinitesimal transformations of the group SU($n$). Here I shall give only the main idea of the proof, and leave the details to the reader.

We know that all unitary matrices $C$ can be reduced to diagonal form

$$C = Q^{-1} \begin{pmatrix} e^{i\alpha} & & \\ & e^{i\beta} & \\ & & \ddots \end{pmatrix} Q$$

$$= Q^{-1} e^{iA} Q = e^{iQ^{-1}AQ} = e^{iB}$$

where $Q$ is unitary, while $A$ and $B$ are self-adjoint. Conversely, every matrix $e^{iB}$, where $B$ is self-adjoint, is unitary. Substituting $C = \exp(iB)$ in (33.5), one obtains an equation which holds for all self-adjoint matrices $B$. It follows that it holds for all $B$, and from this we may conclude that (33.7) holds identically in the $c_l = c_{\lambda'\lambda}$. This method of proof was called by H. Weyl the "unitary trick". The idea of this trick is, to pass from the representation theory of the SL($n$) to the representation theory of SU($n$) and conversely.

Another example of the application of the Unitary Trick. It is known that all continuous representations of SU($n$) are unitary and hence completely reducible. From this, Weyl deduces that all continuous representations of SL($n$) are completely reducible. See H. Weyl: Math. Zeitschr. **23** (1925), p. 271.

We now return to our tensor space $\mathscr{W}$, the space of all multilinear forms

$$\Sigma \, c_{\lambda\mu\ldots\nu} u_\lambda v_\mu \ldots w_\nu.$$

It does not matter whether we consider the case $n = 2$ (the case of spin functions of the electrons in an atom) or the general case; it also does not matter whether we consider the unitary group SU($n$) or the special or general linear group SL($n$) or GL($n$). In any case, we have a representation $\pi$ of the Permutation Group $\mathscr{S}_f$, and a representation $\delta$ of the group SU($n$) or SL($n$) or GL($n$). The Main Theorem says that the set $\sigma$ of all matrices permuting with the matrices of $\pi$ is just the set of all linear combinations of the matrices of $\delta$. Hence, if any subspace of $\mathscr{V}$ is invariant under $\sigma$, it is also invariant under $\delta$ and conversely. If a subspace is irreducible under $\sigma$, it is also irreducible under $\delta$. If two subspaces are equivalent with respect to $\sigma$, they are also equivalent with respect to $\delta$, and conversely.

Making use of the theorems of § 15, it is easy to construct the set $\sigma$. The representation $\pi$ of $\mathscr{S}_f$ is completely reducible, hence the space $\mathscr{W}$ is a direct sum of irreducible subspaces. Let the vectors

$$v_{11}, \ldots, v_{1m}$$

form the basis of such a subspace $\mathscr{V}_1$, and let

$$v_{21}, \ldots, v_{2m}$$

$$\ldots\ldots\ldots\ldots\ldots$$

$$v_{k1}, \ldots, v_{km}$$

be the corresponding basis vectors of the irreducible subspaces $\mathscr{V}_2, \ldots, \mathscr{V}_k$ equivalent to $\mathscr{V}_1$. Thus, we obtain a rectangle of basis vectors

(33.8)
$$\begin{array}{l} v_{11}, \ldots, v_{1m} \\ \vdots \\ v_{k1}, \ldots, v_{km} \end{array}$$

such that the rows of the rectangle are transformed under $\mathscr{S}_f$ by one and the same irreducible representation $\varDelta'$ of $\mathscr{S}_f$. Now, according to § 15, the set of transformations $\sigma$ commuting with $\pi$ transforms any column of the rectangle (33.8) in a quite arbitrary way, and the other columns in just the same way. The same thing holds for the other rectangles.

It follows that every column of a rectangle like (33.8) defines an irreducible subspace of $\mathscr{W}$ with respect to $\sigma$ and hence also with respect to $\delta$. The other columns of the same rectangle (33.8) define equivalent irreducible subspaces. The same thing holds for the other rectangles. Every rectangle thus gives rise to just one irreducible representation of $SU(n)$ or $SL(n)$ or $GL(n)$, as the case may be, and different rectangles give rise to inequivalent representations. Hence there are just as many inequivalent representations $\varDelta'$ of $\mathscr{S}_f$ contained in $\pi$ as there are inequivalent irreducible representations $\varrho'$ of $SU(n)$ or $SL(n)$ or $GL(n)$ contained in $\delta$. To every $\varDelta'$ corresponds just one rectangle and just one $\varrho'$.

In the special case of $SU(2)$ the representations $\varrho'$ are just the representations $\varrho_S$ of degree $2S+1$ known from § 19 and § 20. Since we have just $f$ electrons, each of them having spin $\frac{1}{2}$, the total spin $S$ has one of the values

$$S = \frac{f}{2} - g$$

where $g$ is an integer less than or equal to $\frac{1}{2} f$. Hence we have the result:

*To every value $S = \frac{1}{2} f - g$ corresponds just one representation $\Delta'_S$ of $\mathscr{S}_f$ contained in the representation $\pi$, and to different values of $S$ correspond inequivalent representations $\Delta'_S$.*

We shall now write down the vectors $v_{ij}$ forming the rectangles explicitly. For this purpose, we have to change the notation a little. Instead of $u, v, ..., w$ we shall now use the letters

$$u, v, w, ..., p, q, r, s, t$$

to denote pairs of covariant variables such as $u_1, u_2$, etc. up to $t_1, t_2$. By $x_1, x_2$ we shall denote a contravariant pair of variables, transformed under SU(2) or SL(2) in such a way that

$$u_1 x_1 + u_2 x_2$$

remains invariant. From all these variables we shall form a polynomial domain

$$\mathbb{C}(u_1, u_2; v_1, v_2; ... ; t_1, t_2; x_1, x_2).$$

In analogy to § 21, we now form the invariant product

(33.9)
$$B = (u_1 v_2 - u_2 v_1) ... (p_1 q_2 - p_2 q_1)$$
$$\cdot (r_1 x_1 + r_2 x_2) ... (t_1 x_1 + t_2 x_2)$$

formed from $g$ invariant factors like $u_1 v_2 - u_2 v_2$ and from $f - 2g$ invariant factors like $r_1 x_1 + r_2 x_2$. The number $g$ between 0 and $\frac{1}{2} f$ may be chosen at will; the difference $f - 2g$ is just $2S$. In the expansion of (33.9) the monomials

$$X_S^M = \frac{x_1^{S+M} x_2^{S-M}}{\sqrt{(S+M)! (S-M)!}}$$

are multiplied by certain coefficients $W_S^M$, which are multilinear expressions in the $u, v, ..., r, ..., t$. Just as in § 21, one proves that the $W_S^M$ are transformed under SU(2) according to the representation $\varrho_S$. Hence we can take the $X_S^M$ as one of the columns of a rectangle. The other columns of the same rectangle are obtained from the first column by permutations of the letters $u, v, ..., t$. One goes on permuting the letters as long as one obtains linearly independent vectors in any one of the rows of the rectangle.

**Example.** Let $f$ be 3. In the 8-dimensional space generated by the products $u_\lambda v_\mu w_\nu$ one can form the following rectangles

$$S = \tfrac{3}{2}: \quad \boxed{\begin{array}{l} \sqrt{3} \cdot u_1 v_1 w_1 \\ u_1 v_1 w_2 + u_1 v_2 w_1 + u_2 v_1 w_1 \\ u_1 v_2 w_2 + u_2 v_1 w_2 + u_2 v_2 w_1 \\ \sqrt{3} \cdot u_2 v_2 w_2 \end{array}}$$

$$S = \tfrac{1}{2}: \quad \boxed{\begin{array}{ll} (u_1 v_2 - u_2 v_1) w_1 & (u_1 v_2 - u_2 v_1) v_1 \\ (u_1 v_2 - u_2 v_1) w_2 & (u_1 w_2 - u_2 w_1) v_1 \end{array}}$$

The spin functions $W_S^M$, the coefficients of the expression (33.9), can be characterized by the fact that they are antisymmetric in the first $g$ pairs of electrons and symmetric in the other $f - 2g$ electrons. In the vector diagram the spin vectors of the first $g$ pairs annihilate each other (for instance $u_1 v_2$ is a spin function, in which the first spin is directed upwards, the second downwards), whereas the remaining $f - 2g$ electrons have parallel spins (directed upwards in the product $r_1 s_1 \ldots t_1$, and down in $r_2 s_2 \ldots t_2$). Thus, the total spin is

$$S = \tfrac{1}{2}(f - 2g) = \tfrac{1}{2} f - g,$$

as it should be according to the general theory.

Our construction of the irreducible representations $\Delta'_S$ implies that the matrix elements of these representations are rational numbers. The same thing holds not only for the representations $\Delta'_S$, but for *all* irreducible representations $\Delta'$ of the symmetric group, as we shall see in the next § 34. It follows that the irreducible representations of $\mathscr{S}_f$ are equal to their conjugate complex representations $\Delta'^*$. Now the representations $\Delta'$ are all unitary, hence the conjugate complex representations $\Delta'^*$ are equivalent to the dual representations $\widetilde{\Delta'}$. Hence $\Delta'$ is equivalent to $\widetilde{\Delta'}$, and the relation

$$\Delta \times A \sim \widetilde{\Delta'}$$

derived in § 31 simplifies to

$$\Delta \times A \sim \Delta'$$

which can also be written as (33.1):

$$(33.10) \qquad\qquad \varDelta \sim \varDelta' \times A \,.$$

## § 34. Representations of the Symmetric Group $\mathscr{S}_n$

This section is reprinted from my "Algebra" II, § 110. The proofs given here are all due to John von Neumann. The theory was first developed by G. Frobenius (1903).

We consider the group $\mathscr{S}_n$ of permutations of $n$ digits $1, 2, ..., n$ and seek its absolutely irreducible representations in, say, the field $\Omega$ of all algebraic numbers. It will turn out that these representations are actually rational, that is, are found in the field $\mathbb{Q}$ of rational numbers.

We start with the group ring $\mathscr{R} = s_1 \Omega + \cdots + s_{n!}\Omega$ and consider its left ideals. Every such left ideal is a direct sum of minimal left ideals; these ideals provide the irreducible representations. Since each left ideal is generated by an idempotent element, we first look for the idempotent elements.

We write the numerals $1, 2, ..., n$ in any order in $h$ successive rows ($h$ arbitrary) so that in the $v$ th row there are $\alpha_v$ numbers and the conditions

$$(34.1) \qquad \begin{cases} \alpha_1 \geqq \alpha_2 \geqq \cdots \geqq \alpha_h \\ \sum_{v=1}^{h} \alpha_v = n \end{cases}$$

are satisfied. We write the first elements of the $h$ rows all under one another, likewise the second elements, and so on, for example, as in the following schema in which the points represent numerals:

$$\vdots \vdots \cdot \qquad (\alpha_1, \alpha_2, \alpha_3) = (3, 2, 2); \qquad n = 7 \,.$$

Such an arrangement of the numerals $1, 2, ..., n$ we call a *schema* $\Sigma_\alpha$. The index $\alpha$ denotes the sequence $(\alpha_1, \alpha_2, ..., \alpha_h)$. The possible indices are ordered by the following convention: $\alpha > \beta$ if the first nonvanishing difference $\alpha_v - \beta_v$ is positive. For example, in the case $n = 5$,

$$(5) > (4, 1) > (3, 2) > (3, 1, 1) > (2, 2, 1) > (2, 1, 1, 1) > (1, 1, 1, 1, 1) \,.$$

Given a schema $\Sigma_\alpha$, we denote by $p$ all those permutations which permute the numerals within the rows of a schema but leave the rows themselves invariant; similarly, we denote by $q$ those permutations which permute only the numerals within the columns of a schema. For each fixed $q$ the symbol $\sigma_q$ denotes the number $+1$ or $-1$ according to whether $q$ is an even or odd permutation. If $s$ is any permutation we denote by $s\,\Sigma_\alpha$ the schema into which $\Sigma_\alpha$ is transformed by the permutation $s$. It is easily seen that if $q$ leaves the columns of $\Sigma_\alpha$ invariant, then $sqs^{-1}$ leaves the columns of $s\,\Sigma_\alpha$ invariant, and conversely. Finally, we put (in the group ring $\mathscr{R}$), for each fixed $\Sigma_\alpha$,

$$S_\alpha = \sum_p p$$

$$A_\alpha = \sum_q q\sigma_q.$$

The following rules are easily verified:

(34.2) $$pS_\alpha = S_\alpha p = S_\alpha$$

(34.3) $$A_\alpha q\sigma_q = qA_\alpha\sigma_q = A_\alpha.$$

From (34.2) and (34.3) it now follows that $S_\alpha$ and $A_\alpha$ are idempotent up to a factor $f_\alpha$. The additional algebraic properties of $S_\alpha$ and $A_\alpha$ follow from the following *combinatorial lemma*.

**Lemma.** *Let $\Sigma_\alpha$ and $\Sigma_\beta$ be two schemata of the above type, and let $\alpha \geq \beta$. If in $\Sigma_\alpha$ there are nowhere two numerals in a single row which occur in $\Sigma_\beta$ in the same column, then $\alpha = \beta$ and the schema $\Sigma_\alpha$ can be transformed by a permutation of the form $pq$ into the schema $\Sigma_\beta$:*

$$pq\,\Sigma_\alpha = \Sigma_\beta.$$

(Here $p$ and $q$ refer to $\Sigma_\alpha$; that is, $p$ leaves the rows and $q$ the columns of $\Sigma_\alpha$ invariant.)

*Proof.* $\alpha \geq \beta$ implies $\alpha_1 \geq \beta_1$. In the first row of $\Sigma_\alpha$ there are $\alpha_1$ numerals. If these same numerals in $\Sigma_\beta$ are all in distinct columns, then $\Sigma_\beta$ must have at least $\alpha_1$ columns; from this it follows that $\alpha_1 \leq \beta_1$ and hence $\alpha_1 = \beta_1$. These numerals can all be brought into the first row of $\Sigma_\beta$ by a permutation $q'_1$ which leaves the columns of $\Sigma_\beta$ invariant.

Further, $\alpha \geq \beta$ implies $\alpha_2 \geq \beta_2$. In the second row of $\Sigma_\alpha$ there are $\alpha_2$ numerals. If these are all in distinct columns in $q'_1\,\Sigma_\beta$, then, apart from the first row, $q'_1\,\Sigma_\beta$ must still have at least $\alpha_2$ columns. This implies that $\alpha_2 \leq \beta_2$ and hence $\alpha_2 = \beta_2$. These numerals can all be brought into the second row of $\Sigma_\beta$ by a permutation which leaves both the columns of $q'_1\,\Sigma_\beta$ and the first row invariant.

Continuing in this manner, we finally obtain a schema $q' \Sigma_\beta = q'_h \ldots q'_2 q'_1 \Sigma_\beta$ whose rows coincide with those of $\Sigma_\alpha$. Therefore $\Sigma_\alpha$ can be transformed into $q' \Sigma_\beta$ by a permutation $p$:

$$q' \Sigma_\beta = p \Sigma_\alpha.$$

The permutation $q' = q'_h \ldots q'_2 q'_1$ leaves the columns of $\Sigma_\beta$ invariant; it therefore also leaves the columns of $q' \Sigma_\beta = p \Sigma_\alpha$ invariant. For appropriate $q$, then,

$$q' = pq^{-1}p^{-1}$$

and hence

$$pq^{-1}p^{-1} \Sigma_\beta = p \Sigma_\alpha$$

$$\Sigma_\beta = pq \Sigma_\alpha, \quad \text{Q.E.D.}$$

The combinatorial lemma implies first of all that

(34.4)                    $A_\beta S_\alpha = 0 \quad \text{for} \quad a > \beta.$

For by the lemma if $\alpha > \beta$ there must exist a pair of numerals which occur in a single row of $\Sigma_\alpha$ and in a single column of $\Sigma_\beta$. If $t$ is the transposition which interchanges the numerals of this pair, then, by (34.2) and (34.3),

$$A_\beta S_\alpha = A_\beta t t^{-1} S_\alpha = - A_\beta S_\alpha,$$

which gives (34.4).

Similarly,

$$S_\alpha A_\beta = 0 \quad \text{for} \quad \alpha > \beta.$$

Now all transforms of $A_\beta$ are also annihilated by $S_\alpha$:

$$S_\alpha s A_\beta s^{-1} = 0 \quad \text{for} \quad \alpha > \beta;$$

since $s A_\beta s^{-1}$ is again an $A_\beta$ which belongs to the permuted schema $s \Sigma_\beta$. On multiplying by $s\Omega$ and summing over all $s$ in $\mathscr{G}$, this result implies

$$S_\alpha (\Sigma \, s\Omega) \, A_\beta = (0)$$

or

(34.5)                    $S_\alpha \mathscr{R} A_\beta = (0) \quad (\alpha > \beta).$

The left ideals $\mathscr{R} A_\beta$ with $\beta < \alpha$ are therefore annihilated by $S_\alpha$; this means that $S_\alpha$ is represented by zero in the representation provided by $\mathscr{R} A_\beta$. On the other hand $S_\alpha A_\alpha \neq 0$, since the coefficient of the identity element in the product $S_\alpha A_\alpha$ does not vanish. Therefore $S_\alpha$ is not represented by zero in the representation given by $\mathscr{R} A_\alpha$; this representation thus contains at least one irreducible component which occurs in no $\mathscr{R} A_\beta$ with $\beta < \alpha$. This irreducible component we shall now determine more explicitly.

The element $S_\alpha A_\alpha = \sum_p \sum_q pq\sigma_q$ has, by (34.2) and (34.3), the property

$$pS_\alpha A_\alpha q\sigma_q = S_\alpha A_\alpha.$$

We now prove that up to a factor, $S_\alpha A_\alpha$ is the only element with this property. We show: *if an element a of $\mathscr{R}$ has the property*

$$(34.6) \qquad\qquad paq\sigma_q = a$$

*for all p and q, then a must have the form* $(S_\alpha A_\alpha) \cdot \gamma$.

*Proof.* We put

$$(34.7) \qquad\qquad a = \sum_s s\gamma_s \quad (\gamma_s \in \Omega).$$

Substituting (34.7) in (34.6) gives:

$$(34.8) \qquad\qquad \sum_s s\gamma_s = \sum_s psq\sigma_q\gamma_s.$$

On the left side, only one term with $pq$ occurs, namely $pq\gamma_{pq}$; on the right side there is also only one term containing $pq$, namely the term with $s = 1$. Equating coefficients gives

$$\gamma_{pq} = \sigma_q\gamma_1.$$

We now select an $s$ which does not have the form $pq$. Then $s\Sigma_\alpha$ is distinct from all the $pq\Sigma_\alpha$ and thus by the combinatorial lemma there are two numerals $i, j$ which occur in $\Sigma_\alpha$ in a single row and also in $s\Sigma_\alpha$ in a single column. If $t$ is the transposition of these two numerals, $t = (jk)$, then $t' = s^{-1}ts$ interchanges only the numerals $s^{-1}j$ and $s^{-1}k$ which appear in the same column in $s^{-1}s\Sigma_\alpha = \Sigma_\alpha$. Therefore $t$ is a permutation of type $p$, and $t'$ a permutation of type $q$. In (34.8) we may therefore put $p = t$ and $q = t'$; for this special $s$, then,

$$psq = tss^{-1}ts = s$$

$$\sigma_q = -1;$$

comparison of the terms with $s$ on the left and right in (34.8) gives

$$\gamma_s = -\gamma_s, \qquad \gamma_s = 0.$$

In (34.7), therefore, only terms with $s = pq$, $\gamma_s = \sigma_q \gamma_1$ occur, and hence

$$a = \sum_{p,q} pq\sigma_q \gamma_1 = (S_\alpha A_\alpha)\gamma_1, \qquad \text{Q.E.D.}$$

From what has just been proved it follows immediately that for every element $b$ of $\mathscr{R}$ the element $S_\alpha b A_\alpha$ has the form $(S_\alpha A_\alpha)\gamma$, since, for each $p$ and each $q$,

$$pS_\alpha b A_\alpha q\sigma_q = S_\alpha b A_\alpha.$$

Thus

$$S_\alpha \mathscr{R} A_\alpha \subseteq (S_\alpha A_\alpha)\Omega.$$

Putting $S_\alpha A_\alpha = I_\alpha$, it follows that

(34.9)                    $$I_\alpha \mathscr{R} I_\alpha \subseteq S_\alpha \mathscr{R} A_\alpha \subseteq I_\alpha \Omega.$$

We now assert that $\mathscr{R}I_\alpha$ is a minimal left ideal. Indeed, if $l$ is a subideal of $\mathscr{R}I_\alpha$, then it follows from (34.9) that

$$I_\alpha l \subseteq I_\alpha \Omega,$$

and hence, since $I_\alpha \Omega$ is a minimal $\Omega$-module, either

$$I_\alpha l = I_\alpha \Omega \quad \text{or} \quad I_\alpha l = (0).$$

In the first case it follows that $\mathscr{R}I_\alpha = \mathscr{R}I_\alpha \Omega \subseteq \mathscr{R}I_\alpha l \subseteq l$, and hence $l = \mathscr{R}I_\alpha$. In the second case it follows that $l^2 \subseteq \mathscr{R}I_\alpha l = (0)$, and hence $l = (0)$, since there are no nilpotent ideals except (0).

The minimal left ideals $\mathscr{R}I_\alpha$ and $\mathscr{R}I_\beta$ are not operator-isomorphic for $\alpha > \beta$. For from (34.5), for $\alpha > \beta$,

$$S_\alpha \mathscr{R} I_\beta = S_\alpha \mathscr{R} S_\beta A_\beta \subseteq S_\alpha \mathscr{R} A_\beta = (0),$$

and hence, for each $a'$ of $\mathscr{R}I_\beta$,

$$S_\alpha a' = 0.$$

If now $\mathscr{R}I_\alpha \cong \mathscr{R}I_\beta$, then it would follow that, for each $a$ of $\mathscr{R}I_\alpha$,

$$S_\alpha a = 0 ;$$

this, however, is not true for $a = I_\alpha = S_\alpha A_\alpha$, since $S_\alpha^2 A_\alpha = f_\alpha S_\alpha A_\alpha \neq 0$.

Each left ideal $\mathscr{R}I_\alpha$ provides an irreducible representation $\varrho_\alpha$, and these representations are inequivalent for distinct $\alpha$ by the above remarks.

The number of representations $\varrho_\alpha$ thus found is equal to the number of solutions of (34.1). This number is at the same time the number of classes of conjugate permutations; for each such class consists of all elements which decompose into cycles of definite lengths $\alpha_1, \alpha_2, \ldots, \alpha_h$, and these lengths can be ordered in accordance with the conditions (34.1). However, since the number of *all* inequivalent irreducible representations is given by the number of classes of conjugate permutations, it follows that *up to equivalence the representations $\varrho_\alpha$ exhaust all irreducible representations of the symmetric group $\mathscr{S}_n$.*

Chapter VI

# Molecule Spectra

## § 35. The Quantum Numbers of the Molecule

In a first approximation, a molecule may be considered as a system of
$f$ electrons moving around nuclei fixed in space. We shall restrict ourselves to molecules consisting of two atoms, and we shall neglect the
spin. We may assume the nuclei to be situated on the $Z$-axis in distances
$\beta'r$ and $\beta''r$ from their common centre of gravity, where $r$ is the distance
between the nuclei and

$$\beta' = \frac{M''}{M' + M''}, \qquad \beta'' = \frac{M'}{M' + M''},$$

$M'$ and $M''$ being the masses of the nuclei.

The eigenvalue problem of the electrons in the field of the two fixed
nuclei is invariant under the group $\mathscr{R}_2$ of axial rotations and reflexions.
In § 12 (Example 3) we have determined the representations of this
group $\mathscr{R}_2$. The results may be summarized as follows: Eigenfunctions
occur in pairs $(\varphi_\Lambda, \varphi_{-\Lambda})$, where $\Lambda = 0, 1, 2,...$ is the *axial quantum number*.
Under rotations about the $Z$-axis over an angle $\gamma$ the function $\varphi_\Lambda$ is
multiplied by $e^{-i\Lambda\gamma}$, and $\varphi_{-\Lambda}$ by $e^{i\Lambda\gamma}$. In the case $\Lambda = 0$ there are two
kinds of eigenfunctions $\varphi_0^+$ and $\varphi_0^-$, which are multiplied by $+1$ and $-1$
under a reflection $s_y$. In these cases we shall write $\Lambda = 0^+$ or $\Lambda = 0^-$.
The corresponding representations of the group $\mathscr{R}_2$ are called $\varrho_0^+$
and $\varrho_0^-$. In the case $\Lambda > 0$ we have two eigenfunctions $\varphi_\Lambda$ and $\varphi_{-\Lambda}$,
belonging to one and the same energy value, which are interchanged by
the reflection $s_y$. This representation is called $\varrho_\Lambda$. The representations
$\varrho_0^+$, $\varrho_0^-$ and $\varrho_\Lambda(\Lambda = 1, 2,...)$ are the only possible irreducible one-valued
representations of the group $A$.

The energy terms belonging to the values $\Lambda = 0^+, 0^-, 1, 2, 3,...$ are
denoted by Greek letters $\Sigma^+, \Sigma^-, \Pi, \Delta, \Phi,...$ corresponding to the Latin
letters $S, P, D, F,...$ used for atomic terms.

For an infinitesimal rotation $I_z$ about the $Z$-axis we have

$$I_z\varphi_\Lambda = -i\Lambda\varphi_\Lambda$$

hence

$$L_z \varphi_A = i I_z \varphi_A = \Lambda \varphi_A.$$

This means: In the state $\varphi_A$ the component $\hbar L_z$ of the angular momentum has the eigenvalue $\hbar \Lambda$.

In reality the molecule is not a system having fixed nuclei. The nuclei are in motion, just as the electrons. The system consists of $f + 2$ moving particles. As we have seen in § 5, the number of moving points can be reduced to $f + 1$ by taking the center of gravity as a new origin of coordinates. The moving points are: the $f$ electrons, having coordinates $q_1, ..., q_f$ with respect to the new origin of coordinates, and a "fictitious nucleus", lying at distance $r$ from the new origin in the direction of the line joining the actual nuclei. The coordinates of this fictitious nucleus will henceforth be called $q_0$.

The Schrödinger equation for the new coordinates without spin is equation (5.7) of § 5, in which the third term on the left may be neglected in first approximation, since it is small as compared with the preceding term. This equation is invariant with respect to rotations and reflections leaving fixed the origin of coordinates, i.e. the center of gravity. The eigenfunctions at any fixed energy level form a vector space, which is a direct sum of irreducible subspaces with respect to rotations and reflections. Let us consider one of these subspaces. Under rotations it is transformed according to an irreducible representation $\varrho_K$, where $K$ is an integer. The subspace is generated by $2K + 1$ linearly independent functions

$$\psi^{(m)}(q_0, q_1, ..., q_f) \quad (m = K, K - 1, ..., - K).$$

Moreover, we may suppose that under the inversion

$$x' = -x, \quad y' = -y, \quad z' = -z$$

the functions $\psi$ are multiplied by $w = \pm 1$.

From now on we shall denote by $r$ the variable distance between the nuclei in the problem of $f + 2$ variable mass points, and by $\varrho$ a fixed value of $r$ used in the calculation of the eigenfunction $\varphi$ of the electrons moving around fixed nuclei.

The questions we have to answer are:

1. What is the relation between the eigenfunctions $\varphi_{\pm A}$ obtained for fixed nuclei and the eigenfunctions $\psi^{(m)}$ of our system of $f + 1$ variable points?

2. What is the relation between the quantum number $\Lambda$ of the functions $\varphi_{\pm A}$ and the quantum numbers $K, m, w$ of the functions $\psi^{(m)}$?

3. What is the relation between the energy value $E(\varrho)$ found for a molecule having fixed nuclei at distance $\varrho$, and the energy values obtained from the Schrödinger equation of the $f+2$ freely variable mass points?

A rotation $R$ transforms the function $\psi^{(m)}$ into a new function $'\psi^{(m)}$ defined by

(35.1)                     $'\psi^{(m)}(Rq_0,\ldots,Rq_f)=\psi^{(m)}(q_0,\ldots,q_f)\,.$

The new function $'\psi^{(m)}$ can be expressed as a linear combination of the original functions $\psi^{(m)}$ as follows:

(35.2)                     $'\psi^{(m)}(q)=\sum_g a_{gm}(R)\,\psi^{(g)}(q)\,,$

the $a_{gm}(R)$ being the matrix elements of the matrix representing $R$ in the representation $\varrho_K$. The letter $q$ stands for the arguments $q_0,\ldots,q_f$. For these arguments we may substitute whatever we like, for (35.2) is an identity in the $q$. Substituting the arguments $Rq_0,\ldots,Rq_f$, one obtains, because of (35.1),

(35.3)          $\psi^{(m)}(q_1,\ldots,q_f)=\Sigma\, a_{gm}(R)\,\psi^{(g)}(Rq_0,\ldots,Rq_f)\,.$

For $R$ we may take any rotation we like. We may choose for $R$ a rotation transforming the point $q_0$ into a point $Q$ with coordinates $(0,0,r)$ on the $Z$-axis. Of course, $R$ is not uniquely determined by the point $q_0$. If $R$ is chosen in an arbitrary way (depending, of course, on the situation of the point $q_0$), $R$ may afterwards be replaced by $R_\gamma R$, where $R_\gamma$ is a rotation about the $Z$-axis over an angle $\gamma$.

Let us choose an $R$ transforming $q_0$ into $Q$, and substitute it into (35.3). Since $Rq_0=Q$, we obtain

(35.4)          $\psi^{(m)}(q_0,\ldots,q_f)=\sum_g a_{gm}(R)\,\psi^{(g)}(Q,Rq_1,\ldots,Rq_f)\,.$

Since $Q=(0,0,r)$ depends on one coordinate $r$ only, the functions

(35.5)               $\psi_Q^{(g)}(q_1,\ldots,q_f)=\psi^{(g)}(Q,q_1,\ldots,q_f)$

contain two variables less than the original functions $\psi^{(m)}(q_0,\ldots,q_f)$. Substituting (35.5) into (35.4), one obtains the fundamental formula

(35.6)          $\psi^{(m)}(q_0,\ldots,q_f)=\sum_g a_{gm}(R)\,\psi_Q^{(g)}(Rq_1,\ldots,Rq_f)\,.$

By this formula, the problem of finding a set of functions $\psi^{(m)}$ transforming under rotations according to the representation $\varrho_K$ is reduced to the simpler problem of finding a set of functions $\psi_Q^{(g)}$ containing two variables less.

The right side of (35.6) depends on the point $Q$ and on the rotation $R$ which transforms $q_0$ into $Q$. The left side depends only on $q_0 = R^{-1}Q$. Hence the functions $\psi_Q^{(g)}$ on the right cannot be quite arbitrary: they must satisfy a condition, which can be derived as follows.

In (35.6) we may take for $R$ a rotation $R_\gamma$ over an angle $\gamma$ about the $Z$-axis, which transforms $Q$ into itself. The matrix $A(R_\gamma)$ representing $R_\gamma$ in the representation $\varrho_K$ is a diagonal matrix with elements

$$a_{gg}(R_\gamma) = e^{-ig\gamma},$$

hence (35.6) becomes

$$\psi_Q^{(g)}(q_1,\ldots,q_f) = e^{-ig\gamma}\psi_Q^{(g)}(R_\gamma q_1,\ldots,R_\gamma q_f)$$

or

$$R_\gamma \psi_Q^{(g)}(R_\gamma q_1,\ldots,R_\gamma q_f) = e^{-ig\gamma}\psi_Q^{(g)}(R_\gamma q_1,\ldots,R_\gamma q_f)$$

or, if we replace $R_\gamma q_1,\ldots,R_\gamma q_f$ by $q_1,\ldots,q_f$:

$$(35.7) \qquad\qquad R_\gamma \psi_Q^{(g)} = e^{-ig\gamma}\psi_Q^{(g)}.$$

This condition (35.7) is sufficient. Indeed, if one takes for $\psi_Q^{(g)}$ in (35.3) arbitrary functions of $Q$ and $q_1,\ldots,q_f$ satisfying (35.7) and calculates the right side of (35.6), the resulting functions $\psi^{(m)}$ do not depend on $R$ but only on $q_0 = R^{-1}Q$. For if one replaces, on the right side of (35.6), the rotation $R$ by $R_\gamma R$, one obtains

$$\sum_g e^{-ig\gamma}a_{gm}(R)\,\psi_Q^{(g)}(R_\gamma R q_1,\ldots,R_\gamma R q_f) = \Sigma\, a_{gm}(R)\,\psi_Q^{(g)}(R q_1,\ldots,R q_f)$$

i.e. the function $\psi^{(m)}$ defined by (35.6) remains unchanged.

We shall now study the behaviour of the functions $\psi_Q^{(g)}$ under a reflection $s_y$:

$$x' = x, \quad y' = -y, \quad z' = z.$$

We may obtain this reflection as a product $R_y s$, where $R_y$ is a rotation over an angle $\pi$ about the $y$-axis:

$$x' = -x, \quad y' = y, \quad z' = -z$$

whereas $s$ is the inversion

$$x' = -x, \quad y' = -y, \quad z' = -z.$$

In order to calculate $\psi^{(m)}(sq_0,\ldots,sq_f)$ according to our fundamental formula (35.6), we have to find a rotation which transforms $sq_0$ into $Q$. Such a rotation is $R_yR$, for $R$ transforms $sq_0$ into $sQ$, and $R_y$ transforms $sQ$ into $Q$. Hence we have

$$\psi^{(m)}(sq_0,\ldots,sq_f) = \sum_g a_{gm}(R_yR)\,\psi_Q^{(g)}(R_yRsq_1,\ldots,R_yRsq_f)\,.$$

Now $Rs = sR$ and $R_y s = s_y$, hence

$$(35.8) \qquad \psi^{(m)}(sq_0,\ldots,sq_f) = \sum_h \sum_g a_{gh}(R_y)\,a_{hm}(R)\,\psi_Q^{(g)}(s_yRq_1,\ldots,s_yRq_f)\,.$$

In order to calculate $a_{gh}(R_y)$ we have to investigate how the basis vectors

$$v_g = \frac{u_1^{K+g}u_2^{K-g}}{\sqrt{(K+g)!\,(K-g)!}}$$

of the representation $\varrho_K$ are transformed under $R_y$. Now $R_y$ transforms

$$u_1^{K+g}u_2^{K-g} \quad \text{into} \quad (-1)^{K-g}u_1^{K-g}u_2^{K+g}$$

(see § 19), hence we have

$$a_{gh}(R_y) = (-1)^{K-g} \quad \text{if} \quad h = -g\,,$$
$$= 0 \quad \text{otherwise}\,.$$

Substituting this into (35.8), we obtain

$$(35.9)\quad \psi^{(m)}(sq_0,\ldots,sq_f) = \sum_h (-1)^{K+h}\,a_{hm}(R)\,\psi_Q^{(-h)}(s_yRq_1,\ldots,s_yRq_f)\,.$$

We have supposed that the functions $\psi^{(m)}$ have the inversion character $w$. This is the case, if the expression (35.9) is equal to $w\cdot\psi^{(m)}$, i.e. if

$$(-1)^{K+h}\,\psi_Q^{(-h)}(s_yRq_1,\ldots,s_yRq_f) = w\cdot\psi_Q^{(h)}(Rq_1,\ldots,Rq_f)$$

or, which is equivalent,

$$(-1)^{K+g}\,\psi_Q^{(-g)}(s_yq_1,\ldots,s_yq_f) = w\cdot\psi_Q^{(g)}(q_1,\ldots,q_f)$$

or, still simpler,

$$(35.10) \qquad\qquad s_y\psi_Q^{(g)} = (-1)^{K+g}\,w\psi_Q^{(-g)}\,.$$

The simplest way to satisfy the conditions (35.7) and (35.10) is, to assume that only two of the functions $\psi_Q^{(g)}$, namely $\psi_Q^{(\Lambda)}$ and $\psi_Q^{(-\Lambda)}$, are different from zero, or if $g = 0$ only one function $\psi^{(0)}$. In the first case $(\Lambda > 0)$ we have a pair of functions

$$\psi_Q^{(\Lambda)}, \psi_Q^{(-\Lambda)}$$

which are transformed under axial rotations and reflections according to the representation $\varrho_\Lambda$. In the second case we have one function $\psi_Q^{(0)}$ which takes under a reflection $s_y$ a factor $(-1)^K w$, and we have the representation

$$\varrho_0^+ \quad \text{if} \quad (-1)^K = w,$$

$$\varrho_0^- \quad \text{if} \quad (-1)^K = -w.$$

In § 36 we shall prove that the functions $\psi_Q^{(g)}$ may be assumed, in a fairly good approximation, to have the form

(35.11) $$\psi_Q^{(g)} = f(r)\, \varphi_g(r, q_1, \ldots, q_f)\,.$$

The first factor $f(r)$ determines the vibrations of the distance $r$ about its equilibrium value: $f(r)$ is an eigenfunction of the differential equation of a non-harmonic oscillator. The second factor $\varphi_g (g = \pm \Lambda)$ determines the motion of the electrons about the two nuclei, which are supposed to be fixed in space.

The physical significance of the quantum numbers $\Lambda$ and $K$ is easy to find. As we have seen, $\varphi_\Lambda$ is an eigenfunction of the operator $L_z$ with eigenvalue $\Lambda$. Hence the operator $\hbar L_z$, the component of the angular momentum $\hbar L$ of the electrons in the direction of the axis connecting the nuclei, has the value $\hbar \Lambda$ in the state $\varphi_\Lambda$ (and of course the value $-\hbar \Lambda$ in the state $\varphi_{-\Lambda}$).

On the other hand, $\hbar K$ is the *total* angular momentum of the molecule. In any one of the states $\psi^{(m)}$, the operator $\mathscr{L}^2$ has the eigenvalue $K(K+1)$, and the component $\hbar L_z$ of $\hbar \mathscr{L}$ has the eigenvalue $\hbar m$, where $m$ takes on all integer values from $K$ to $-K$. The quantum number $K$, which determines the total angular momentum, is called the *rotation quantum number* of the molecule.

Since $g = \pm \Lambda$ can only assume integer values between $K$ and $-K$, we have

$$K \geqq \Lambda\,.$$

It follows that for a given value of $\Lambda$, the rotation quantum number can only assume the values

$$K = \Lambda, \Lambda + 1, \Lambda + 2, \ldots.$$

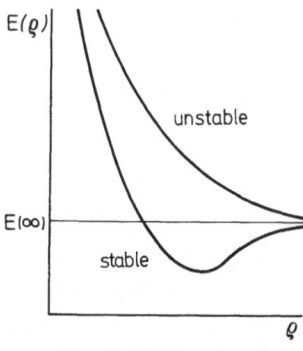

Fig. 10. The function $E(\varrho)$

Now, how can one determine all possible states of the molecule?

First one keeps the nuclei fixed, in a reasonable distance $\varrho$, and determines the possible states $\varphi_g$ of the electrons. They occur in pairs $(\varphi_A, \varphi_{-A})$ having the same energy. The functions $\varphi_A$ and $\varphi_{-A}$ are eigenfunctions of a differential equation which will be derived in § 36. The energy of any state is a function $E(\varrho)$ of $\varrho$. For stable molecules this function has a minimum, and the actual values of $\varrho$ are near this minimum (see Figure 10).

As soon as the function $E(\varrho)$ is known, the eigenvalue problem for the function $f(r)$ determining the vibrations can be solved (see § 36). The solutions depend on a *vibration quantum number* $v = 1, 2, 3,....$ They also depend on the rotation quantum number $K$, but the energy-values belonging to different values of $K$ do not differ very much. Hence we get, for every $\varLambda$ and every $v$, a *band* consisting of energy levels lying close together. The singel levels of each band are distinguished by their rotation quantum numbers $K = \varLambda, \varLambda + 1, \varLambda + 2$, etc.

From general group-theoretical considerations we have derived selection rules for the atomic quantum numbers $L$ and $w$. Of course, the same rules hold for $K$ and $w$:

$$(35.12) \qquad \begin{cases} K \to K - 1, K, K + 1 \text{ (except } 0 \to 0), \\ w \to -w. \end{cases}$$

To derive the selection rule for $\varLambda$, we consider the matrix elements of the electrical momentum of the electrons and nuclei taken together. It is easy to see that the contribution of the nuclei is very small because of their slow motion, so we may restrict ourselves to the electrons. We have to multiply the operators $X = \Sigma e x_\nu$, $Y, Z$ with the eigenfunctions (35.6) and to express the resulting products as sums of eigenfunctions (35.6).

The result is an identity in $q_0$, which must hold in particular for $q_0 = Q$. In this case we can take for $R$ the identity, so that (35.6) reduces to

$$\psi^{(m)}(Q, q_1, \ldots, q_f) = \psi_Q^{(g)}(q_1, \ldots, q_f) \quad \text{if} \quad m = g = \pm \Lambda ,$$

$$= 0 \quad \text{otherwise.}$$

Let us consider the case $g = + \Lambda$; the case $g = - \Lambda$ can be treated just so. Under rotations $R_y$ the function $\psi_Q^{(\Lambda)}$ is multiplied by $e^{-i\Lambda y}$, and the functions

$$(X + iY)\psi_Q^{(\Lambda)}, \quad (X - iY)\psi_Q^{(\Lambda)}, \quad Z\psi_Q^{(\Lambda)}$$

are multiplied by

$$e^{-i(\Lambda+1)y}, \quad e^{-i(\Lambda-1)y}, \quad e^{-i\Lambda y}$$

respectively. Hence, in the expansion of these products, only terms $\varphi_Q^{\Lambda'}$ with

$$\Lambda' = \Lambda + 1, \Lambda \quad \text{or} \quad \Lambda - 1$$

can occur. Just so, if $\Lambda$ is $0^+$, the expansion of $(X \pm iY)\psi_Q^{(0)}$ and of $Z\psi_Q^{(0)}$ contains only terms with $\Lambda' = 1$ or $0^+$, and if $\Lambda$ is $0^-$, we find only terms with $\Lambda' = 1$ or $0^-$. Hence the selection rule for $\Lambda$ is:

$$\Lambda \rightarrow \Lambda + 1 \quad \text{or} \quad \Lambda \quad \text{or} \quad \Lambda - 1$$

but not $0^+ \rightarrow 0^-$ and not $0^- \rightarrow 0^+$.

Note that in the cases $0^+ \rightarrow 0^+$ and $0^- \rightarrow 0^-$ the quantum number $K$ must jump to $K - 1$ or $K + 1$, because otherwise the selection rule $w \rightarrow - w$ would be violated.

## § 36. The Rotation Levels

In § 35 we have stated that the functions $\varphi_Q^{(g)}$ may be assumed, in a fairly good approximation, to be products (35.11)

$$\psi_Q^{(g)} = f(r)\, \varphi_g(r, q_1, \ldots, q_f) ,$$

in which $f(r)$ is an eigenfunction of an oscillation problem, whereas the two functions $\varphi_g(g = \pm \Lambda)$ are solutions of the Schrödinger equation of the $f$ electrons moving about fixed nuclei.

To prove these assertions, we go back to the wave equation of the molecule. In § 5 we have shown how one can obtain a wave equation for the relative motion of a set of particles with respect to its center of gravity. The resulting Eq. (5.9) is valid for a molecule as well as for an atom, but the number of particles is now $f + 2$ instead of $f + 1$, so we have to write

$$\left\{ - \sum_1^{f+1} \frac{\hbar^2}{2\mu_\alpha} \Delta'_\alpha + \frac{\hbar^2}{2M} \left( \sum_1^{f+1} \frac{\partial}{\partial q'_\alpha} \right)^2 + U \right\} \psi = E\psi ,$$

the $q'_\alpha$ being the relative coordinates of the second nucleus and the electrons with respect to the center of gravity $q_c$, and $M$ being the total mass of the molecule.

In the second term, the sum from 1 to $f + 1$, multiplied by $(\hbar/i)$, is the operator for the sum of the momenta of the second nucleus and the electrons. Because of the small factor $\hbar^2/2M$, the momenta of the electrons can be neglected, so that the sum may be reduced to just one term $\alpha = 1$, which can be united with the first term in the first sum. Thus we obtain the simplified equation

$$\left( - \frac{\hbar^2}{2\mu_1} + \frac{\hbar^2}{2M} \right) \Delta'_1 - \sum_1^{f+1} \frac{\hbar^2}{2\mu} \Delta'_\alpha + U \right) \psi = E\psi .$$

In the same approximation the center of gravity of the whole molecule can be identified with the center of gravity of the nuclei, and $M$ may be replaced by $\mu_0 + \mu_1$. Instead of the difference

$$q'_1 = q_1 - q_c = q_1 - \frac{\mu_0 q_0 + \mu_1 q_1}{\mu_0 + \mu_1} = \frac{\mu_0}{\mu_0 + \mu_1} (q_1 - q_0)$$

we may introduce the three components of the vector $q^* = q_1 - q_0$ as new coordinates. Thus our equation reduces to

$$\left\{ - \frac{\hbar^2}{2M^*} \Delta^* - \sum_1^{f+1} \frac{\hbar^2}{2\mu} \Delta'_\alpha + U \right\} \psi = E\psi ,$$

$M^*$ being a fictitious mass

$$M^* = \frac{\mu_0 \mu_1}{\mu_0 + \mu_1} .$$

Returning to our previous notation, we may give the electrons numbers from 1 to $f$ and write $q_0$ and $\Delta_0$ instead of $q^*$ and $\Delta^*$, thus obtaining

(36.1) $$\left\{ - \frac{\hbar^2}{2M^*} \Delta_0 - \sum_1^{f} \frac{\hbar^2}{2\mu} \Delta_\alpha + U \right\} \psi = E\psi .$$

Into this fundamental equation we have to substitute for $\psi$ the expression (35.6):

$$(36.2) \qquad \psi^{(m)} = \sum_g a_{gm}(R)\, \psi_Q^{(g)}(Rq_1, \ldots, Rq_f).$$

The main problem is the evaluation of the first term in (36.1). We have to calculate $\Delta_0 \psi^{(m)}$. As in §6, we may introduce polar coordinates $r, \vartheta, \varphi$ for $q_0$ and write

$$\Delta_0 = \frac{\partial^2}{\partial r^2} + \frac{2}{r}\frac{\partial}{\partial r} + \frac{1}{r^2} D_0,$$

where $D_0$ operates only on $\vartheta$ and $\varphi$. The first two terms give no difficulty: they do not operate on $R$, but only on $r$, which means that on the right of (36.2) they operate only on $\psi_Q^{(g)}$. So we are left with the calculation of $D_0 \psi^{(m)}$.

According to 8.6, the operator $D_0$ may be written as

$$D_0 = -\mathscr{L}_0^2 = -(L_{0x}^2 + L_{0y}^2 + L_{0z}^2).$$

A direct evaluation of $-\mathscr{L}_0^2 \psi^{(m)}$ is not easy, because the right side of (36.2) depends on $\vartheta$ and $\varphi$ in a rather complicated way. On the right side of (36.2) a rotation $R$ occurs which carries the point $q_0$ into the north pole $Q = (0, 0, \varrho)$ of the sphere on which $Q$ lies. We may choose for $R$ a rotation over an angle $\vartheta$ around an axis perpendicular to the plane $0Qq_0$, but the differentiation of this rotation with respect to $\vartheta$ and $\varrho$ is not an easy matter.

For this reason, we shall introduce the total angular momentum $\mathscr{L}$ with components

$$L_x = L_{0x} + L_{1x} + \cdots + L_{fx}, \text{ etc.}$$

and the angular momentum $\mathscr{L}'$ of the electrons with components

$$L_x' = L_{1x} + \cdots + L_{fx}, \text{ etc.}$$

We may note that $L_x$ commutes with $L_x'$, just so $L_y$ with $L_y'$, and $L_z$ with $L_z'$. Hence we have

$$(36.3) \qquad -D_0 = \mathscr{L}_0^2 = (\mathscr{L} - \mathscr{L}')^2 = \mathscr{L}^2 - 2\mathscr{L}' \cdot \mathscr{L} + \mathscr{L}'^2.$$

In the state $\psi = \psi^{(m)}$ the total angular momentum of the molecule is $\hbar K$, hence we have

$$\mathscr{L}^2 \psi = K(K+1)\, \psi.$$

The operator $\mathscr{L}'^2$ operates only on the electrons. The product $\varrho^{-2}\mathscr{L}'^2$ has the same order of magnitude as the operators $\varDelta_1,\ldots,\varDelta_f$, but in (36.1) it occurs with a factor $\hbar^2/2M^*$, which is at least one thousand times smaller than the factor $\hbar^2/2\mu$ in the other terms of (36.1). Therefore, in a first approximation, the term $\mathscr{L}'^2$ can be omitted. A closer analysis shows that the middle term $-2\mathscr{L}'\mathscr{L}$ in (36.3) has the same order of magnitude as the last term $\mathscr{L}'^2$, hence both terms may be omitted. In this approximation, (36.1) reduces to

$$(36.4) \qquad \frac{\hbar^2}{2M^*}\left(-\frac{\partial^2}{\partial r^2} - \frac{2}{r}\frac{\partial}{\partial r} + \frac{K(K+1)}{r^2}\right)\psi$$

$$-\frac{\hbar^2}{2\mu}\sum_1^f \varDelta_\alpha\psi + U\psi = E\psi.$$

Now let us return to the fundamental equation (35.6) of § 35:

$$(36.5) \qquad \psi^{(m)}(q_0,\ldots,q_f) = \sum_g a_{gm}(R)\,\psi_Q^{(g)}(Rq_1,\ldots,Rq_f).$$

This function $\psi^{(m)}$ has to satisfy the differential equation (36.4). In this differential equation, differentiations with respect to $\vartheta$ and $\varphi$ do not occur any more: $\vartheta$ and $\varphi$ (and hence $R$) may be regarded as constant. Hence, if the single functions $\psi_Q^{(g)}$ on the right satisfy the differential equation, so do the functions $\psi^{(m)}$ on the left for all values of $m$ from $K$ to $-K$; and conversely. So we may take for $\psi_Q^{(g)}$ any set of $2K+1$ functions satisfying (36.4) and having the right transformation properties (35.7 and (35.10):

$$(36.6) \qquad \begin{cases} R_\gamma\psi_Q^{(g)} = e^{-ig\gamma}\,\psi_Q^{(g)}, \\ s_y\psi_Q^{(g)} = (-1)^{K+g}\,w\psi_Q^{(-g)}. \end{cases}$$

In particular, it is permissible to assume that only two of the functions $\psi_Q^{(g)}$ are different from zero. These two functions

$$\psi_Q^{(\pm A)}(Rq_1,\ldots,Rq_f)$$

have to satisfy (36.4) for any fixed value of $R$. In particular, we may choose for $R$ the identity, thus obtaining the pair of functions

$$\psi_Q^{(\pm A)}(q_1,\ldots,q_f)$$

which have to satisfy the differential equation (36.4). Of course, it is sufficient to consider only one function $\psi_Q^{(A)}$; the other one can be obtained from it by a reflection $s_y$.

The physical signification of (36.4) is as follows. The rotation of the nuclei is eliminated: the nuclei move up and down on the $Z$-axis, their common center of gravity being the origin of coordinates. Each of the nuclei oscillates about its equilibrium position. The "centrifugal force" of the rotation gives rise to a term

(36.7)
$$\frac{\hbar^2}{2M^*} \frac{K(K+1)}{r^2}$$

in the Hamiltonian.

A still better approximation can be obtained by using the method of perturbation theory: the eigenfunctions $\psi^{(m)}$ can be retained, but the energy value may be slightly different from the first approximation obtained from (36.4). The corrected value $E$ may depend on $\Lambda$. In the German edition of this book the correction was obtained in a rough approximation by replacing the term (36.7) by

(36.8)
$$\frac{\hbar^2}{2M^*} \frac{K(K+1)-\Lambda^2}{r^2}.$$

However, I do not know how good this approximation is.

We now try to find an approximate solution of (36.4) by substituting for $\psi$ a product of the form

(36.9)
$$\psi = f(r)\,\varphi(r, q_1,\ldots,q_f),$$

$\varphi$ being a solution of the Schrödinger equation for the motion of $f$ electrons about nuclei fixed in a distance $r$ (or $\varrho$) on the $Z$-axis. This Schrödinger equation reads

(36.10)
$$-\frac{\hbar^2}{2\mu} \sum_1^f \Delta_\alpha \varphi + U\varphi = E(r)\,\varphi.$$

In solving this equation we may replace the variable $r$ by a fixed value $\varrho$, but it should be kept in mind that the solution $\varphi$ of this equation depends not only on $q_1, \ldots q_f$ but also on $r$.

If we substitute $\psi = f(r)\,\varphi$ into (36.4), the operator

$$-\frac{\partial^2}{\partial r^2} - \frac{2}{r}\frac{\partial}{\partial r}$$

must be applied not only to the first factor $f$ but also to the second factor $\varphi$. This gives us terms like

$$-2f_r\varphi_r - f\varphi_{rr} - \frac{2}{r}f\varphi_r$$

which are, however, multiplied by the small factor $\hbar^2/2M^*$. Since $M^*$ is large as compared with $\mu$, these terms may be neglected. Thus, Eq. (36.4) reduces to

$$\frac{\hbar^2}{2M^*}\left(-f_{rr}-\frac{2}{r}f_r+\frac{K(K+1)}{r^2}f\right)\cdot\varphi$$

$$+f(r)\left(-\frac{\hbar^2}{2\mu}\Sigma\,\Delta_\alpha\varphi+U\varphi\right)=Ef\cdot\varphi.$$

Because of (36.10), the second term of the left may be replaced by

$$f(r)\cdot E(r)\,\varphi,$$

so we are left with a differential equation for $f$:

$$(36.11)\quad\left(-\frac{\hbar^2}{2M^*}\left(\frac{\partial^2}{\partial r^2}+\frac{2}{r}\frac{\partial}{\partial r}\right)+E(r)+\frac{\hbar^2}{2M^*}\frac{K(K+1)}{r^2}\right)f=Ef.$$

The Eq. (36.11) has the same form as the Schrödinger equation for vibrations of a mass point moving on a straight line under the influence of a force having potential energy

$$(36.12)\qquad\qquad U(r)=E(r)+\frac{\hbar^2}{2M^*}\frac{K(K+1)}{r^2}.$$

A stable molecule is possible only if the function $U(r)$ has a minimum less than its value at infinity $U(\infty)$. The main term in (36.12), namely $E(r)$, is just the energy of a system consisting of two fixed nuclei in a distance $r$ and $f$ electrons. For $r\to0$ the energy $E(r)$ tends to infinity, and for $r\to\infty$ the energy tends to the energy $U(\infty)$ of two separate atoms or ions (see Fig. 10). The second term on the right in (36.12) is due to the "centrifugal force" arising from the rotation of the nuclei. Once more, a better value of the energy $E$ can be found by perturbation theory, leaving the function $\psi^{(m)}$ unchanged and calculating $\langle H\psi^{(m)},\psi^{(m)}\rangle$ from the original, exact Hamiltonian. However, since the general form and behaviour of the functions $\psi^{(m)}$ is more important than the exact value of $E$, this application of perturbation theory is not very interesting. Moreover, since the neglected terms in the Hamiltonian are of the order of magnitude of $(\mu/M)$ times the not-neglected terms, and since the function $E(r)$ is not accurately known anyhow, a better approximation to the neglected terms seems pointless.

All in all, in order to obtain a useful approximation to the eigenfunctions of the molecule, we have to take the following steps:

*First*, solve the differential eq. (36.10) for the eigenfunctions of the electrons moving about nuclei fixed in a distance $r$. This is the most difficult step. It yields a pair of eigenfunctions $\varphi^{(\pm\Lambda)}$ and an energy function $E(r)$.

*Secondly*, solve the eigenvalue problem (36.11) for the vibrations of the nuclei and form the products

$$\psi_Q^{(g)} = f(r)\,\varphi^{(g)}(r, q_1, \ldots, q_f) \quad (g = \pm\Lambda).$$

*Thirdly*, calculate the eigenfunctions $\psi^{(m)}$ of the molecule by (36.2):

$$\psi^{(m)} = \sum_g a_{gm}(R)\,\psi_Q^{(g)}(Rq_1, \ldots, Rq_f)$$

Now, let us have a closer look at the eigenfunctions $f(r)$ of the vibration problem (36.11) and their eigenvalues $E$. Let $\Lambda$ and $\varphi^{(\Lambda)}$ be kept fixed. If the function $E(r)$ has a minimum less than $E(\infty)$ as in Fig. 10, there may be a finite or infinite number of eigenvalues $E_v (v = 1, 2, 3, \ldots)$. The number $v$ is called the *vibration quantum number*. The energy $E_v$ depends not only on $v$, but also on the *rotation quantum number* $K$, which determines the third term in (36.11); so we may write $E_v(K)$ instead of $E_v$. Now if $K$ takes the values $\Lambda, \Lambda+1, \Lambda+2, \ldots$, the third term in (36.11) does not change very much. As long as $K$ is not too large, the changes in the term

$$\frac{\hbar^2}{2M^*}\frac{K(K+1)}{r^2}$$

from any $K$ to the next $K+1$ are usually small as compared with the changes from one vibration term $E_v$ to the next $E_{v+1}$. Therefore one has, for every fixed value of $v$, a sequence of *rotation levels*

$$E_v(\Lambda),\ E_v(\Lambda+1),\ E_v(\Lambda+2),\ldots$$

lying close together.

Now let us see what consequences this situation has for the emission or absorption spectrum. Suppose the state of the electron configuration jumps from one state $\varphi$ to another state $\varphi'$. At the same time the vibration quantum number $v$ may jump to $v'$. As for $K$, there are several possibilities for $K$ and $K'$, giving rise to absorption or emission lines. All these lines, belonging to different values of $K$ and $K'$, but to one and the same jump of $\varphi$ to $\varphi'$ and of $v$ to $v'$, lie close together and from a *band*. For $K$ we have the selection rule

$$K \to K-1 \quad \text{or} \quad K \quad \text{or} \quad K+1,$$

hence the band consists of three *branches*:

$$the\ P\text{-}branch\quad K \to K+1$$

$$the\ Q\text{-}branch\quad K \to K$$

$$the\ R\text{-}branch\quad K \to K-1$$

The arrows are valid for emission lines; for absorption lines the arrows must be inverted. If the initial and final state are both $\Sigma$-states (i.e. $\Lambda = 0^+$ or $0^-$), the number $K$ must jump to $K+1$ or $K-1$ (see the end of § 35), and the $Q$-branch cannot occur.

If the spin of the electrons is taken into account, new complications arise, which will not be discussed here. I only note that if the spin quantum number is zero for the initial and hence also for the final state, no complications arise: all terms remain singlets. If a spin is present and if the multiplet splitting is small as compared with the distance between successive rotation terms [1], the spin can be treated as a small perturbation, and one obtains a splitting of each term with rotation quantum number $K$ and spin $S$ into a multiplet. Just as in § 27, each term of the multiplet is characterized by a quantum number $J$ which takes the values

$$J = K+S, K+S-1,\dots,|K-S|\,.$$

## § 37. The Case of Two Equal Nuclei

If the two nuclei in the molecule have equal charges, the two-center problem considered in § 35 has another symmetry: it is transformed into itself by the inversion $s$ with respect to the center of gravity:

$$x' = -x, \quad y' = -y, \quad z' = -z\,.$$

The inversion $s$ commutes with all operations of the axial group of rotations and reflections. If the operation $s$ is applied to the eigenfunctions $\varphi(q)$ of the electrons without spin, they are multiplied by a factor $\varepsilon = \pm 1$. The same thing holds for if the spin is taken into account, for the pure spin functions remain invariant under the operation $s$.

We know that the functions $\varphi$ occur in pairs $\varphi_\Lambda, \varphi_{-\Lambda}$ of equal energy. Now if $\varphi_\Lambda$ belongs to a certain value $\varepsilon$, i.e. if

$$s\varphi_\Lambda = \varepsilon\varphi_\Lambda\,,$$

---

[1] This is the case for light molecules such as $H_2$, and for all $\Sigma$-terms.

then $\varphi_{-A} = s_y \varphi_A$ belongs to the same $\varepsilon$, for we have

$$s(s_y \varphi_A) = s_y s \varphi_A = s_y \varepsilon \varphi_A = \varepsilon(s_y \varphi_A).$$

The terms with $\varepsilon = +1$ or $-1$ are usually denoted as follows:

$$\varepsilon = +1 : \Sigma_g, \Pi_g, \Delta_g, \ldots : \text{"even terms"},$$

$$\varepsilon = -1 : \Sigma_u, \Pi_u, \Delta_u, \ldots : \text{"odd terms"}.$$

If the reflection $s$ is multiplied by the rotation over $180°$ about the $Z$-axis (the line connecting the nuclei), wo obtain the reflection with respect to the $XY$-plane:

$$s \cdot R_z = R_z \cdot s = s_z$$

and we see that under this reflection the functions $\varphi_{\pm A}$ are multiplied by $(-1)^A \varepsilon$. In what follows we shall not use this reflection, because the use of the inversion $s$ leads to simpler formulae.

If we suppose that the nuclei of equal charge (or atomic number) also have equal mass, it follows that the differential equation of the freely moving molecule is also invariant if the nuclei are interchanged or, what amounts to the same thing, if $q_0$ is replaced by $-q_0$. The eigenfunction $\psi$ of the whole system may be symmetric or antisymmetric, i.e. it is multiplied by $\chi = +1$ or by $\chi = -1$ if $q_0$ is replaced by $-q_0$. We now shall derive the relation between $\varepsilon$ and $\chi$.

If the permutation of the nuclei $(q_0 \to -q_0)$ and the inversion $s$ of the whole system $(q_0 \to -q_0, q_1 \to -q_1, \text{etc.})$ are combined, we obtain the transformation

$$q_1 \to -q_1, \ldots, q_f \to -q_f$$

i.e. the inversion $s$ applied to the electrons only. Under this transformation, the eigenfunctions are multiplied by $w \cdot \chi$. This holds also if the nuclei are kept fixed, i.e. the functions $\varphi_{\pm A}$ are also multiplied by $w \cdot \chi$, and we have

(37.1) $$\varepsilon = w \cdot \chi$$

For the symmetry character $\chi$ the selection rule $\chi \to \chi$ holds, for if $\psi$ is symmetric or antisymmetric in the coordinates of the nuclei, so are $X\psi, Y\psi$ and $Z\psi$. As in the case of atoms, we also have $w \to -w$, hence the selection rule for $\varepsilon$ reads

$$\varepsilon \to -\varepsilon,$$

i.e.: *Even terms combine with odd terms only, and conversely.*

# Author and Subject Index

# Grundlehren der mathematischen Wissenschaften

*A Series of Comprehensive Studies in Mathematics*

## A Selection